Environmental Remediation

ACS SYMPOSIUM SERIES **509**

Environmental Remediation

Removing Organic and Metal Ion Pollutants

G. F. Vandegrift, EDITOR
Argonne National Laboratory

D. T. Reed, EDITOR
Argonne National Laboratory

I. R. Tasker, EDITOR
National Institute for Petroleum and Energy Research

Developed from a symposium sponsored
by the Division of Industrial and Engineering Chemistry, Inc.,
at the 201st National Meeting
of the American Chemical Society,
Atlanta, Georgia,
April 14–19, 1991

American Chemical Society, Washington, DC 1992

Library of Congress Cataloging-in-Publication Data

Environmental remediation: removing organic and metal ion pollutants /
G. F. Vandegrift, ed., D. T. Reed, ed., I. R. Tasker, ed.

p.   cm.—(ACS Symposium Series, 0097–6156; 509).

"Developed from a symposium sponsored by the Division of Industrial
and Engineering Chemistry, Inc., at the 201st National Meeting of the
American Chemical Society, Atlanta, Georgia, April 14–19, 1991."

Includes bibliographical references and indexes.

ISBN 0–8412–2479–X

1. Water, Underground—Purification—Congresses. 2. Soil
pollution—Congresses. 3. Separation (Technology)—Congresses.
4. Bioremediation—Congresses.

I. Vandegrift, G. F. (George F.), 1945–   . II. Reed, Donald
Timothy, 1956–   . III. Tasker, I. R. (Ian R.), 1954–   . IV. American
Chemical Society. Division of Industrial and Engineering Chemistry.
V. American Chemical Society. Meeting (201st: 1991: Atlanta, Ga.).
VI. Series.

TD426.E59   1992
628.5'.2—dc20   92–27219
                                CIP

TD426
.E59
1992

The paper used in this publication meets the minimum requirements of American National
Standard for Information Sciences—Permanence of Paper for Printed Library Materials, ANSI
Z39.48–1984.   ∞

PRINTED IN THE UNITED STATES OF AMERICA

DEC 12 1994

Foreword

THE ACS SYMPOSIUM SERIES was first published in 1974 to provide a mechanism for publishing symposia quickly in book form. The purpose of this series is to publish comprehensive books developed from symposia, which are usually "snapshots in time" of the current research being done on a topic, plus some review material on the topic. For this reason, it is necessary that the papers be published as quickly as possible.

Before a symposium-based book is put under contract, the proposed table of contents is reviewed for appropriateness to the topic and for comprehensiveness of the collection. Some papers are excluded at this point, and others are added to round out the scope of the volume. In addition, a draft of each paper is peer-reviewed prior to final acceptance or rejection. This anonymous review process is supervised by the organizer(s) of the symposium, who become the editor(s) of the book. The authors then revise their papers according the the recommendations of both the reviewers and the editors, prepare camera-ready copy, and submit the final papers to the editors, who check that all necessary revisions have been made.

As a rule, only original research papers and original review papers are included in the volumes. Verbatim reproductions of previously published papers are not accepted.

M. Joan Comstock
Series Editor

Contents

INDEXES

Preface

THE UNITED STATES HAS SUFFERED severe environmental damage as a result of industrial growth and defense-related activities. Cleanup costs are estimated to be as high as a trillion dollars. Our damage to the environment is already affecting our health and welfare, and it must be repaired to ensure our survival. The U.S. Environmental Protection Agency estimates that more than 75,000 registered hazardous waste generators and more than 25,000 possible hazardous waste sites exist in the United States.

Although the situation is ominous, it has encouraging aspects. Environmental remediation and waste avoidance technologies are rapidly increasing growth areas. Potentially great benefits await those who can develop economical, effective, and efficient solutions to these problems.

The symposium on which this volume is based acknowledged the growing importance of environmental remediation and reflected our belief that separation science will continue to play a key role in the remediation of contaminated aquifers, as well as surface and subsurface media. The wide variety of developing technologies discussed will provide the newly initiated with an understanding of the breadth of the problems and potential solutions and will widen the perspective of those already in the field. The various chapters cover the following topics: improvements and applications of existing separation technology, developing technologies, applications of separation science to waste minimization and preconcentration, basic research applied to understanding the chemistry behind new technologies, and analysis of the hazards present in the environment.

Chapter 1 presents an overview of the extent of our abuse of the environment, the high cost of its cleanup, and the federal regulations that govern cleanup and disposal of present waste. It also summarizes some recently published common-sense suggestions on how we should proceed.

The table of contents illustrates the breadth of this volume. About one-third of the chapters deal directly with remediation technologies for groundwater and soil decontamination; another third deal with waste treatment avoidance technologies; and the last third discuss fundamental research for developing new technologies or for measuring the problem or the effectiveness of the treatment. This book will provide a contribution to this important field and will reemphasize the need for continued progress and development.

We acknowledge the efforts of two extremely capable, amiable, and undauntable ladies and thank them for their hard work, their persistence, and their goodwill throughout this endeavor.

The first is Donna Tipton—who kept our correspondence organized, handled the typing load and much of the telephone communication, and made excuses for us to Anne Wilson—who has been helpful, tough but gracious, and persistent in moving this publication along.

G. F. VANDEGRIFT
Chemical Technology Division
Argonne National Laboratory
Argonne, IL 60439

D. T. REED
Chemical Technology Division
Argonne National Laboratory
Argonne, IL 60439

I. R. TASKER
National Institute for Petroleum
  and Energy Research
IIT Research Institute
P.O. Box 2128
Bartlesville, OK 74005

June 19, 1992

Chapter 1

Environmental Restoration and Separation Science

D. T. Reed[1], I. R. Tasker[2], J. C. Cunnane[1], and G. F. Vandegrift[1]

[1]Chemical Technology Division, Argonne National Laboratory, 9700 South Cass Avenue, Argonne, IL 60439
[2]National Institute for Petroleum and Energy Research, IIT Research Institute, P.O. Box 2128, Bartlesville, OK 74005

The problem of environmental restoration, specifically the cleanup of contaminated soils and groundwaters, is one of the most important technical and societal problems we face today. To provide a background to this problem, the extent and cost, laws and regulations, important contaminants, and key issues in environmental restoration are discussed. A brief introduction to the role of separation science, in relation to environmental restoration, is also given.

The cleanup of anthropogenic contaminants that are present in the environment is one of the most important problems that we face today. These contaminants can cause a wide range of pollution problems, including global climate change, ozone depletion, ecological deterioration, and groundwater contamination. The remediation of existing waste, along with concerns over the fate of waste we are currently generating or plan to generate, is recognized by the public as the leading environmental issue of today *(1,2)*.

The applications of separation science to environmental restoration are centered on the cleanup of contaminated groundwaters and soils. The extent and complexity of the groundwater contamination problem continues to present formidable technological obstacles to cleanup. The most important factors that contribute to this complexity are the large number of contaminated sites, the wide diversity of the contaminants present in those sites, the inherent complexity of the subsurface chemistry of the contaminants, and the difficulty in interpreting existing regulations to establish compliance and properly prioritize site remediation efforts.

Separation science has already had an important role in the cleanup of contaminated groundwater and soil. It has been

0097–6156/92/0509–0001$06.00/0

successfully applied to solve a wide variety of specific
groundwater contamination problems. However, when one takes into
consideration the limitations of these "successes", and considers
the magnitude and complexity of the cleanup problem that we face
today, it is evident that current technology is inadequate. In
this context, the need for greater emphasis on the development of
new and improved technology has never been greater.

By its very nature, separation science and technology draws
upon the knowledge derived from a large number of science and
engineering disciplines. A comprehensive review of this field
would be an enormous task, and we make no claim to such an
objective in preparing this chapter. Rather, it is our objective
to provide background information on subsurface contamination and
introduce the role of separation science. To this end, we discuss
the magnitude and costs associated with the groundwater
contamination problem, the regulatory requirements that drive
remediation efforts, important organic and inorganic contaminants,
issues associated with environment restoration and give a general
listing of separation techniques that may apply to environmental
restoration. To the extent possible, the reader is provided key
references for more detailed discussions of the various aspects of
this important field.

Extent and Cost of Environmental Remediation

The list of sites that need to undergo environmental restoration is
long and rapidly growing. In 1991, the U.S. Environmental
Protection Agency (EPA) (9-5) had estimated that in the United
States there are over 500,000 hazardous chemicals in use today;
over 75,000 registered hazardous waste generators; and over 4500
hazardous waste treatment, storage, and disposal facilities. This
is further complicated by the number of existing hazardous waste
sites. Current estimates, which are acknowledged to be low, are
that there are over 25,000 possible hazardous waste sites (4).
Another 52,000 municipal landfills, 75,000 on-site industrial
landfills, 180,000 sites with underground storage tanks, and 60,000
abandoned mines still need to be evaluated more thoroughly.

There is also increased recognition of the extent of
environmental contamination within the Department of Energy (DOE)
complex (6-8). Radioactive groundwater plumes have been identified
in the subsurface at many DOE facilities. Remediation efforts and
the efforts to develop new and improved technologies have been
greatly emphasized by DOE in the last few years (6). The overall
waste management problem within the DOE is compounded by the
existence of large amounts of radioactive, hazardous, and
radioactive/hazardous mixed waste currently either awaiting final
disposal or requiring processing prior to final disposal. Recent
estimates of these waste inventories have identified 180 waste
streams and the need for the disposal of over 700,000 m^3 of waste
(8). The majority of this waste (>99%) is located at Hanford in
Washington; Rocky Flats in Colorado; Idaho National Engineering
Laboratory in Idaho; Y-12, K-15 and Oak Ridge National Laboratory
in Tennessee; and the Savannah River Plant in North Carolina.

Associated with the large number of potential waste cleanup sites are the rising cost estimates for the environmental restoration of these sites. The methodology by which the cost of cleanup and environmental restoration is determined is itself a controversial issue. Numerous estimates, based on a wide variety of assumptions, can be found in the literature. The trend in cost estimates, however, is quite readily discerned. In the early summer of 1991, the costs generally quoted for environmental restoration were in the $70-250 billion range. By November of the same year (9), Portney put the cost estimate for cleaning up all civilian and military hazardous waste sites, including Resource Conservation and Recovery Act (RCRA) and Superfund mandates, at $420 billion. Just one month later (10,2), this estimate was raised to $750 billion, with the total possibly surpassing $1 trillion. Given that estimated costs of the average Superfund site is in the range of $30-50 million (11), this estimate should continue to increase rapidly as additional sites are evaluated and added to the list of hazardous waste sites.

The very alarming and rapid increase in estimates of the cleanup cost results from a combination of (1) an increased recognition of the extent of contamination,(2) changes in the regulations governing compliance that increase the cleanup "load" and specify more stringent standards, and (3) a growing recognition that we don't have all the answers and that technological breakthroughs are needed (12,13). It is precisely this trend in cost that drives the need for both short and long-term research to develop new, improved and cost-effective remediation technology.

Laws and Regulations Pertaining to Environmental Restoration

In preparing this work, we found that one feature of the societal problems, the legal feature, is of preeminent importance in driving the economic and scientific side of the remediation effort. However, relevant laws and regulations are not well understood even within the scientific community. Treatment and disposal of hazardous waste, radioactive waste, and mixed waste are regulated by a myriad of statutes that address existing contamination problems, the management of existing waste, and the final disposal of this and future waste generated. The most important of these, along with their interrelationship with various agencies, are shown in Figure 1. The governing regulatory system is complex and either directly or indirectly involves a number of applicable (1) federal and state statutes and local ordinances, (2) regulations promulgated by federal, state, and local agencies, (3) executive orders, and (4) agreements between the involved parties.

In this section, it is our intent to give an overview of only the federal policies (5,14-20) that drive the regulation/remediation of hazardous and nuclear waste. This is not to downplay the important, and often predominant, role played by the states and localities in the overall implementation of hazardous waste management. State-specific regulations are however largely based on federal law and are frequently more restrictive than applicable federal regulations. Applicable regulations and

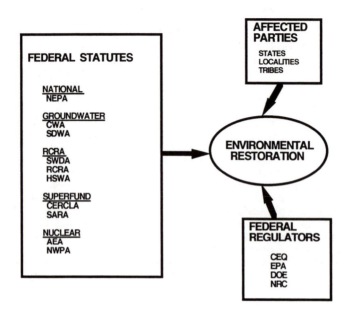

Figure 1. Regulations, statutes and agencies that define environmental restoration.

Definition of Acronyms

AEA   - Atomic Energy Act
CEQ   - Council of Environmental Quality
CERCLA - Comprehensive Environmental Response,
          Compensation & Liability Act
CWA   - Clean Water Act
DOE   - Department of Energy
EPA   - Environmental Protection Agency
HSWA   - Hazardous and Solid Waste Amendment
NEPA   - National Environmental Policy Act
NRC   - Nuclear Regulatory Commission
NWPA   - Nuclear Waste Policy Act
RCRA   - Resource Conservation and Recovery Act
SARA   - Superfund Amendments and Reauthorization Act
SDWA   - Safe Drinking Water Act
SWDA   - Solid Waste Disposal Act

site-specific interpretations of these regulations are usually decided on a case-by-case basis.

Regulation of Hazardous Waste. The first broad-reaching legislation that brought attention to environmental protection was the National Environmental Policy Act (NEPA - 1969). This act provided the basic national charter for environmental protection that recognized a balance between environmental protection and other factors important to national welfare. Its primary objectives were to (1) prevent environmental damage and (2) ensure that all government-agency decision making took environmental factors into account. New activities required the writing of environmental impact statements and, in effect, made environmental protection the mandate of all government agencies.

NEPA established the Council on Environmental Quality (CEQ) as the agency responsible for oversight of other federal agencies. The EPA was designated as a co-participant on this council and has been empowered by subsequent legislation (primarily the Clean Air Act - 1970) to act as the implementing arm of NEPA. EPA is currently chartered to review the environmental impact statements of other agencies.

Even with NEPA, it was not until the mid 1970s that the problem of groundwater contamination/remediation began to receive a significant amount of attention by the EPA. This occurred primarily due to the passage of three series of laws that related to (1) groundwater protection, (2) hazardous waste management, and (3) waste remediation. These acts combined to bring considerable attention to this important problem of groundwater contamination and greatly increased the importance of managing hazardous wastes.

The protection of groundwaters was first addressed in the enactment of the Federal Water Pollution Control Act (FWPCA) in 1972. This brought groundwater controls under the jurisdiction of the federal government. This was amended to focus on the control of toxic pollutants in groundwater and renamed the Clean Water Act (CWA) in 1977. It has been further amended in 1987 to tighten the discharge standards for toxic pollutants. This act currently provides the mechanism by which water quality standards are established. Along the same lines of groundwater protection, the Safe Drinking Water Act (SDWA), passed in 1974, provided for the safety of public water systems and required the EPA to set national drinking water standards. This was amended in 1986 to quicken the pace of its implementation.

The second series of legislation defined hazardous waste management. The Solid Waste Disposal Act (SWDA), enacted in 1965, was the first act passed to regulate waste on a national scale. This was amended by the Resource Conservation and Recovery Act (RCRA), enacted in 1976, and further amended by the Hazardous and Solid Waste Amendments (HSWA) of 1984. These acts collectively provide "cradle to grave" regulation of hazardous waste. Included in these acts are guidelines for management of solid hazardous waste, provisions for strong federal enforcement of the regulations, a regulatory definition of hazardous waste (from which a priority pollutant list was first derived) in 40 CFR 261.2, a detailed regulatory strategy to address hazardous waste management, an identification of financial responsibility for existing waste, and provisions to advance waste-management techniques.

The third series of acts, commonly referred to as
CERCLA/superfund acts, provided the federal government with the
authority to respond to (i.e., clean up) uncontrolled release of
hazardous waste. The Comprehensive Environmental Response,
Compensation, and Liability Act (CERCLA) was enacted in 1977.
CERCLA applies to the release or threat of release into the
environment of any hazardous substance. The broad scope of this
statute is indicated by the fact that "environmental" includes all
environmental media, and "hazardous substance" is broadly defined
to include not only RCRA "hazardous waste" but a list of substances
identified by EPA in 40 CFR 302, which now includes over 700
hazardous substances and over 1500 radionuclides. Whenever there
is a release or threat of a release of a hazardous substance to the
environment, EPA is authorized by the statute to undertake
"removal" and/or "remedial" action.

CERCLA was amended in 1986 by the Superfund Amendments and Reauthorization Act (SARA). This changed the cleanup approach,
increased the involvement of the public in cleanup, and established
a cleanup fund for superfund sites. The enactment of CERCLA/SARA
has spawned a number of related acts that address the notification
of the affected community and response plans. The most important
of these are the Emergency Planning and Community Right-To-Know Act
(EPCRA) and the National Contingency Plan (NCP) enacted in 1986 and
1990, respectively.

Regulation of Radioactive and Radioactive/Hazardous Mixed Waste.
The guidelines for the management of waste containing radioactive
isotopes were established by the Atomic Energy Act (AEA). Under
the provisions of this act, commercial radioactive waste (e.g.,
spent nuclear fuel and low-level radioactive medical waste) was
regulated by the Nuclear Regulatory Commission and codified in
10 CFR part 60 through 71. Radioactive waste generated by the
defense industry, in contrast, was regulated by DOE orders.

DOE's current environmental restoration policy is to clean up
contaminated facilities and sites within the weapons complex to
achieve full compliance with the letter and intent of the
applicable federal, state, and local statutes *(21)*. The Five Year
Plan for Environmental Restoration and Waste Management *(6)*
describes the technologies and research plans currently identified.
Long-term research plans in support of subsurface remediation *(7)*
have also been published and are currently being implemented.

Hazardous waste that contains radioactive material (i.e.,
mixed waste) is regulated under both the AEA and RCRA. Under the
AEA, the EPA has responsibility for setting radiation protection
standards, which are implemented through DOE orders, e.g., Order
5820.2A for DOE radioactive materials *(22)*. This dual regulation
further complicates environmental restoration activities that
involve mixed waste.

Regulatory Approach to Site Remediation. Most environmental
cleanup standards are derived from the provisions of CERCLA,
section 121 "Cleanup Standards" or RCRA, Subtitle C entitled
"Hazardous Waste Management." The implementing regulations are

found, respectively, in 40 CFR part 300 and in 40 CFR parts 264, 265, and 268. Although CERCLA is intended to deal with cleanup of past environmental problems and RCRA is largely intended to prevent future contamination, both statutes and their implementing regulations can affect environmental restoration.

Although the principal statutes and regulations that apply to environmental restoration activities are cited above, standards that derive from other environmental statutes such as CWA, SDWA, NEPA, and state laws still apply. In general, state standards may be substituted for federal standards when the state standards impose requirements that are at least as stringent as the federal standards. Detailed information on the statutes and the associated implementing regulations can be obtained from the "Environmental Guidance Program Reference Books" prepared by Oak Ridge National Laboratory (16). In addition, an overview of the system of environmental statutes and regulations that govern environmental restoration can be obtained from the reference books "Environmental Law Handbook," "Environmental Statutes," and "State Environmental Law Handbooks" published by Government Institutes, Inc. (17-19).

The NCP contains several criteria that are intended to guide decisions on the standards to be achieved in individual remedial actions. Among these the most important are the "threshold criteria," which include (1) a general requirement to protect human health and the environment and (2) cleanup standards which have applicable or relevant and appropriate requirements (ARAR). Under the ARAR approach, EPA can use standards from other federal and state statutes (e.g., CWA, SDWA, RCRA) on a case-by-case basis when these requirements are "applicable or relevant and appropriate." For example, RCRA land-disposal restorations (LDR) may be "relevant and applicable" if a CERCLA remedial action involves RCRA hazardous waste and the waste or its hazardous residue is to be land disposed. In this case, the RCRA LDR standards that are based on the best demonstrated available technology (BDAT) may apply.

Environmental restoration activities may be conducted under a RCRA, Part B permit when RCRA hazardous wastes are involved. The RCRA hazardous wastes are identified in 40 CFR 261 and include "characteristic" hazardous wastes as defined in subpart C and "listed" hazardous wastes as defined in subpart D. The Hazardous and Solid Waste Amendment (HSWA) to RCRA includes prohibitions on land disposal of hazardous waste. Under this statute, the EPA has issued regulations (40 CFR 286) that ban the land disposal of untreated hazardous waste and has established treatment standards based on the BDAT. The way that these standards can be involved in a CERCLA remedial action was discussed above. In addition, technical standards for environmental restoration activities conducted under a RCRA, Part B permit are given in 40 CFR 264, including closure requirements and groundwater concentration limits (see 40 CFR 264.94).

Organic and Metal Pollutants of Importance to Environmental Restoration

The list of metals, radionuclides, and organic compounds that are now recognized as environmental pollutants continues to grow. The

list of 129 priority pollutants identified in RCRA originally
included 13 metals, with the remainder being organic compounds.
When DOE sites are also considered, an additional several hundred
radioisotopes are added to this list.   The list of organics has
grown; over 1000 are now subject to reporting requirements. Any of
the estimated 500,000 organics currently in use have the potential
to be designated as a hazardous material.
     Numerous lists of priority pollutants exist.   It is important
to note that both concentration limits and the list of
priority/regulated contaminants are undergoing constant review and
are often superseded by local and state regulations.   In this
context, we have not tried to tabulate a comprehensive list in this
section. Rather, it is our objective to simply identify some of
the more important contaminants.   This is important from the
perspective of determining the proper emphasis in the development
of separations technology.

Regulatory Definition of Hazardous, Radioactive, and Mixed Waste.
Existing federal regulations give specific regulatory definitions
for all waste types.   Wastes that are of most interest to
environmental restoration and waste management are: hazardous
waste, radioactive waste, and mixed waste.
     Hazardous waste is defined in 40 CFR part 261.3 as solid waste
(as defined in 40 CFR part 261.2) that (1) is not excluded from
regulation as a hazardous waste under section 262.4 and (2) could
cause or significantly contribute to an increase in mortality or an
increase in serious irreversible or incapacitating reversible
illness, or (3) could pose a substantial present or potential
hazard to human health and/or the environment when improperly
stored or treated.
     Materials that are not solid waste and, hence, not subject to
regulation as hazardous waste are domestic sewage, industrial
wastewater that qualifies as point-source discharges (section 402
of the Clean Water Act as amended), irrigation return flows,
nuclear material as defined by the Atomic Energy Act of 1954 (42
U.S.C. 2011 amendment), materials subjected to *in situ* mining
techniques, pulping liquors, spent sulfuric acid, reclaimed
secondary materials, spent wood preserving solution, and the
byproducts of producing coke and coal tar in the steel industry.
     Additionally, all solid waste is not hazardous waste. Solid
waste excluded are household waste, solid waste generated in
farming and raising animals, mining overburden returned to the mine
site, waste generated primarily from the combustion of coal and
fossil fuels, drilling fluids/wastes associated with
oil/gas/geothermal drilling, chromium-containing waste if the waste
generator can demonstrate that it will fail the toxicity
characteristic.
     Operationally, a waste is classified as hazardous based on
three criteria: (1) it is listed as a hazardous waste (40 CFR 261
subpart D); (2) it has one of the following four characteristics:
ignitability, corrosivity, reactivity, and toxicity (see 40 CFR 261
subpart C for the specific definition of these characteristics); or
(3) it falls into the category of "other" hazardous waste
(primarily mixtures of non-hazardous materials with hazardous
waste).

The second type of waste is radioactive waste. This is nonhazardous waste that is categorized primarily according to transuranic content and origin into three classifications: low-level waste, transuranic (TRU) waste, and high-level waste (HLW). Low-level and TRU waste are defined in DOE order 5820.2A (see section II.3a). Low level waste is radioactive waste that contains less than 100 nCi/g of waste of alpha-emitting transuranics with a >20 year half-life. TRU waste is radioactive waste that (1) is not high-level waste and (2) has a specific activity of >100 nCi/g of waste containing transuranic, long-lived alpha emitters. High-level waste is defined in 10 CFR part 60.2 as either (1) irradiated reactor fuel, (2) liquid waste generated from the first cycle of solvent extraction in the processing of nuclear fuel and subsequent concentrates, or (3) solids into which such liquids have been converted.

There is a growing recognition that much of the radioactive waste at DOE sites *(7, 23)* co-exists with hazardous waste that is primarily organic in nature. Waste that contains both radioactive and RCRA-defined hazardous components is classified as mixed waste. This type of waste is subject to both RCRA and DOE/NRC control, whichever is the more stringent.

Metal Contaminants in the Environment. From the perspective of environmental remediation, the focus of separation science should be on the metals currently being regulated from the standpoint of groundwater protection. This is not a hard and fast rule, as there are a number of situations (e.g., specific spills, waste-stream-specific toxins, etc.) where the focus will be on the removal of a specific hazardous metal compound or substance for which standards have not been set. For example, there is currently much emphasis within the DOE on uranium contamination at the Fernald Site in Ohio.

The metals (excluding radionuclides) currently identified for regulation under RCRA/SDWA are listed in Table I. The original list of 13 metals was defined in the 1986 revision of the SDWA. These include two group II metals (barium and beryllium), eight transition metals (cadmium, chromium, copper, lead, mercury, nickel, silver, and thallium), and three near-transition metals (selenium, arsenic, and antimony).

The maximum concentration limits in Table I are currently under review, and lower limits have already been proposed for some metals. In addition to this change, an additional six metals have been proposed for consideration. These, as of the January 1991 revision, are aluminum, manganese, molybdenum, strontium, vanadium, and zinc. Promulgation of this list is in progress, with groundwater protection guidelines expected to be established in the near future.

The importance of the maximum concentration limits established to protect groundwaters in assessing the need for remediation is not entirely clear. CERCLA/superfund cleanup activities are not required to achieve these defined groundwater protection limits. Other factors, such as availability of technology, background groundwater levels, and risk to health, also need to be considered. The trend, however, appears to be in the direction of using the groundwater protection limits as the target and standard for environmental restoration.

Table I.   Metals Listed under RCRA/SDWA as Priority Pollutants

Metals[a]	Toxicity Limits[b] (mg/L)	EPA Guidelines[c] (μg/L)
Arsenic	5.0	5
Barium	100	
Cadmium	1.0	10
Chromium	5.0	100
Lead	5.0	50
Mercury	0.2	2
Selenium	1.0	
Silver	5.0	
Antimony		6
Beryllium		4
Copper		1300
Nickel		100
Thallium		2

[a]Original list proposed in the 1986 amendment to the Safe Drinking Water Act.
[b]Maximum concentration limits currently defined in 40 CFR part 261.24 based on the toxicity criterion.
[c]Reference 23, Table 7 and "Drinking Water Limits Set for 23 More Compounds," Chem. Eng. News, 1992, *70(14)*, 19.

Radionuclides in the Environment.   The 1986 amendment of SDWA also addressed the presence of radionuclides in groundwater from the perspective of radiotoxicity.   Table II lists radionuclides currently regulated under EPA/DOE guidelines.

Table II.   Radioisotopes Currently Regulated under
           EPA/DOE Guidelines

Radionuclide	Concentration Guideline
Beta particle and photon radioactivity[a]	<20 μCi over background
Gross alpha particle activity[a]	15 pCi/L
Radium-226 and -228[a]	5 pCi/L
Uranium[b]	500-600 pCi/L
Strontium[b]	1000 pCi/L
Plutonium[b]	300-400 pCi/L
Cesium[b]	3000 pCi/L

[a]40 CFR part 192.
[b]Reference 24, based on dose-to-man of 100 mRem/year calculations specific to the Hanford site.

Radium and uranium are associated with the mining and/or presence of uranium deposits in the subsurface. Regulations pertaining to these are found in 40 CFR part 192. The other radionuclides are associated with nuclear waste. Groundwater contamination limits for these are defined by dose-to-man calculations from the perspective of radiotoxicity.

There are no isotope or radionuclide-specific concentration standards for the transuranics from the perspective of toxicity. By inventory, the following are the radioactive metals that are of most concern:

Strontium-90
Technetium-99
Tin-126
Cesium-135, -137
Radium-226
Thorium-230, -232
Uranium-233, -234, -235, -236, -238
Neptunium-237
Plutonium-238, -239, -240, -242

All are long-lived radionuclides and are currently receiving attention, to varying extents, within the DOE as subsurface contaminants that may require remediation. Some of these isotopes appear as both toxic and radiotoxic pollutants. In this event, the more stringent criterion applies.

The vast majority of groundwater and subsurface media contaminated with radionuclides are at sites associated with DOE facilities. Although numerous site-specific reports exist, it has only been recently that a general study of the levels of contamination at all DOE sites was made by Zachara and Riley (23). The major observations of this study are the predominance of organics and radionuclide mixtures at existing DOE waste sites. The most important radionuclides identified, based on frequency of appearance in reported analytical results, were plutonium, americium, uranium, neptunium, and cobalt.

Organic Species in the Environment. The predominance of organic species as contaminants in subsurface media is readily recognized (25-30). They have been described as "the most common health-threatening chemicals detected in groundwater," and the greatest difficulties in groundwater remediation have been encountered at organic contamination sites (25).

The scientific and technological problems posed by organic contamination range from microscopic considerations, such as the actual difficulty of bringing about a separation or transformation of a given pollutant in a given environment, to macroscopic considerations such as the huge costs, volumes, and range of materials involved.

Specific lists that identify priority pollutants have been established. The list of regulated organics (SDWA, 1986) is given in Table III. This list includes 14 volatile organics (primarily halogenated hydrocarbons and benzene), 5 microbial species, and 41 non-volatile organics (pesticides and higher molecular weight solvents). The SDWA priority list for future consideration includes an additional 19 pesticides and 43 synthetic organics.

RCRA-related controls exist on over 1000 organic species.
Approximately 200 of these are currently listed as acutely
hazardous waste, with an additional 350 listed as hazardous waste.
A total of 250 organics have been identified for consideration in
establishing groundwater monitoring priority lists (Appendix IX
constituents). Groundwater standards are typically in the low μg/L
range (e.g., 5 μg/L for dichloromethane, 2 μg/L for Endrin, 20 μg/L
for diquat and 1 μg/L for hexachlorobenzene).

Table III. Organic Contaminants Required to Be Regulated
under the SDWA of 1986

Volatile Organics

Trichloroethylene	Benzene
Tetrachloroethylene	Chlorobenzene
Carbon tetrachloride	Dichlorobenzene
1,1,1-Trichloroethane	Trichlorobenzene
1,2-Dichloroethane	1,1-Dichloroethylene
Vinyl chloride	Trans-1,2-Dichloroethylene
Methylene chloride	Cis-1,2-Dichloroethylene

Microbiology and Turbidity

Total coliforms	Viruses
Turbidity	Standard plate count
Giardia lamblia	Legionella

Organics

Endrin	1,1,2-Trichloroethane
Lindane	Vydate
Methoxychlor	Simazine
Toxaphene	PAHs
2,4-D	PCBs
2,4,5-TP	Atrazine
Aldicarb	Phthalates
Chlordane	Acrylamide
Dalapon	Dibromochloropropane (DBCP)
Diquat	1,2-Dichloropropane
Endothall	Pentachlorophenol
Glyphosate	Pichloram.
Carbofuran	Dinoseb
Alachlor	Ethylene dibromide (EDB)
Epichlorohydrin	
Toluene	Xylene
Adipates	Hexachlorocyclopentadiene
2,3,7,8-TCDD(Dioxin)	
Aldicarb sulfone	
Aldicarb sulfoxide	
Ethybenzene	
Heptachlor	
Heptachlor expoxide	
Styrene	

SOURCE: Keith, L. H.; Telliard, W. A., *Environ. Sci.
Technol.*, 1979, *13(4)*, 416.

Issues Associated with Environment Restoration

There are a number of issues associated with remediation of contaminated sites that are currently under debate in the technical community. The most important of these, the impracticality of total remediation, was stated as early as 1980 by the U.S. Geological Survey *(31)* in the following way: "deterioration in [groundwater] quality constitutes a permanent loss of water resource because treatment of the water or rehabilitation of the aquifers is presently generally impractical." This point has been re-stated in more recent commentaries *(2,12,13)*, and the issues of feasibility, cost, contaminant, prioritization, and overall approach to remediation remain. This important debate will continue as more is learned about both the limitations and successes of remediation technology.

Although the list of specific issues is long, a few general issues become clear when the many factors associated with groundwater remediation are considered. The most important of these are:

Existing Contaminated Sites

- Complete and total remediation of all existing groundwater, although a laudable goal, is not realistic. This is due to the enormity of the problem, the limited resources available for the task, and in many instances, the absence of suitable and reliable technology to treat the problem to the extent specified by existing regulations.

- It is important to prioritize existing groundwater problems in terms of health risk/benefit. This is a complex issue that includes re-examination of the regulations that drive and define remediation and re-interpretation of the guidelines in terms of cost and general risk.

- Improved characterization and detection methodology is needed. The vast majority of monitoring is still done by variations on water sampling combined with extensive analysis. Early detection of problems and early detection technology at new waste disposal sites can greatly reduce the extent of contamination.

Future Waste Generated

- Improved waste minimization, documentation, and handling are needed. This is self-evident, with numerous examples of "past sins" in the literature when this was not done.

- Long-term solutions that meet regulatory criteria are needed for waste storage. There are currently no

licensed facilities for the long-term disposal of
hazardous mixed waste, high-level nuclear waste, or
transuranic waste.

Role of Separation Science in Environment Restoration

Treatments available to environmental remediation fall into
three broad categories. These are transformation (i.e.,
destruction), separation, and immobilization. While destruction
and immobilization are ends in and of themselves, separation is
not. In fact, separation only makes sense if it aids in attaining
one or both of these more permanent solutions. For example,
destruction of a hydrocarbon by incineration is feasible if it
existed as a 10 wt % aqueous sludge rather than a 100 ppb aqueous
solution. Final disposal of 10 kg of plutonium would be less
costly and more certain than the cleanup of an equivalent amount of
plutonium in 10^9 kg of plutonium-contaminated soil.

A successful separation process should have two products: a
low-volume stream containing the contaminant(s) in a concentrated
form and a high-volume stream containing the decontaminated matrix.
In general, the concentration of the contaminant(s) is the easier
task. Achieving regulated concentration limits often calls for
very high decontamination levels in the larger-volume,
decontaminated product. This, in some cases, necessitates the
removal of 99.9999% of the contaminant. Also, process design
requirements for concentrating the contaminant(s) and
decontaminating the matrix are generally in opposition, and a
compromise is struck between these two goals. Because a separation
is not an end in itself, the benefits (i.e., goals) of a specific
separation are clearly defined by the disposition of its product
streams.

Based on the old adage, "a stitch in time saves nine," waste
avoidance is an area where separations can be of great benefit to
the environment. The use of separation technologies to
reclaim/recycle materials has become a very important area in
environment restoration, and zero discharge is quickly becoming the
goal of all waste producers. As the cost of the disposal of
hazardous waste increases, industry has come to recognize that
recycle, once more of a public relations effort, is now an economic
necessity.

Lastly, the extremely important role of separations in the
analysis of environmental samples should not be overlooked. The
interplay of technical, societal, and legal factors has driven the
lower limits of detectability and concern to ever decreasing
concentrations. The many separations challenges that exist in this
area have been previously noted in government reports (6, 7, 32).

Assessment of Available Technology. The vital role of separations
and the incentives in probing its frontiers have been well
recognized at the national level by two reports in recent years
(32,33). Although few references focus on separations per se,
there are in the literature numerous on the whole range of
remediation technologies (30,34-41). In addition to this, an
excellent source of information is the literature from the EPA
(42,43).

In a compendium of technologies used in the treatment of hazardous waste (42), technologies are categorized into physical treatment, chemical treatment, biological processes, thermal destruction, and fixation/stabilization processes. Separation technologies are contained entirely within the physical treatment processes section. Those technologies addressed are:

- sedimentation
- centrifugation
- flocculation
- oil/water separation
- dissolved air flotation
- heavy media separation
- evaporation
- air stripping
- steam stripping

- distillation
- soil flushing/washing
- chelation
- liquid/liquid extraction
- supercritical extraction
- filtration
- carbon adsorption
- reverse osmosis
- ion exchange
- electrodialysis

In a collection of synopses of federal demonstrations of innovative site remediation technologies (49), technologies are categorized into bioremediation, chemical treatment, thermal treatment, vapor extraction, soil washing, solidification/stabilization, and other physical treatments. Here separation technologies are contained in the thermal treatment, vapor extraction, soil washing, and other physical treatment sections. Key technologies addressed are:

- desorption and vapor extraction
- low-temperature thermal stripping
- low-temperature thermal treatment
- radiofrequency thermal soil decontamination
- X*TRAX low-temperature thermal desorption
- groundwater vapor recovery system
- in situ stripping with horizontal wells
- in situ soil venting
- in situ steam/air stripping
- integrated vapor extraction and steam vacuum stripping
- Terra Vac in situ vacuum extraction
- vacuum induced soil venting
- BEST solvent extraction
- Biogenesis soil cleaning system
- Biotrol soil washing system
- debris washing system
- Ghea Associates process
- soil treatment with Extraksol
- solvent extraction
- Carver-Greenfield process for extraction of oily waste
- Chemtect gaseous waste treatment
- freeze separation
- membrane microfiltration
- precipitation, microfiltration, and sludge dewatering
- rotary air stripping
- ultrafiltration

A number of these technologies are represented in the contributions to the symposium proceedings that follow this introductory chapter. As reflected in these contributions, there are clear benefits to almost all of these technologies when a suitable environmental restoration problem exists.

Conclusions and Recommendations

The estimated costs of environmental remediation are huge. Regulatory constraints and interpretations are constantly evolving. The technical challenges are staggering. Innovative separation technologies are needed that are economical, acceptable to the public, and effective.

In developing new separations for environmental remediation or waste avoidance, it is important to never disconnect the separation operation from the goals of the entire process. These are:

- Decontamination of the bulk of the groundwater, soil, or waste stream
- Concentration of the contaminant(s)
- Assured and economic final disposal or recycle of the contaminant(s)

Separation is not an end in itself but a means to an end. The disposition of all effluents must be accounted for, and final disposal of equipment must be planned. In general, the products of waste treatment of environmental remediation have no value, and the governing criteria in successful treatment is cost minimization. The separation that produces the least expensive disposal of products at the lowest processing cost is the best process. Social and regulatory pressures will continue to govern what the most important problems are; economics will continue to drive the choices in treatments and long-term research emphasis.

Work supported by the U.S. Department of Energy under Contract W-31-109-Eng-38.

Literature Cited

1. "Evaluation of Groundwater Extraction Remedies," Office of Solid Waste and Engineering Response, United States Environmental Protection Agency, Washington DC, 1989, EPA/540/2-89/054.
2. Abelson, P. H., "Remediation of Hazardous Waste Sites," *Sci.*, 1992, *255*, 901.
3. "The Eighteenth and Nineteenth Annual Report of the Council on Environmental Quality together with the President's Message to Congress," Council on Environmental Quality, U.S. Department of Commerce, N.T.I.S., 1989, PB90-163148.
4. "Characterization of Municipal Solid Waste in the United States, 1960 to 2000 (Update 1988)," Franklin Associates Ltd., U.S. Environmental Protection Agency, Office of Solid Waste, 1988, cited in reference 3, p. 42.
5. Cooke, S. M., *The Law of Hazardous Waste*; Mathew Bender and Co. Inc.: New York, NY, 1991; Vol. 1-3.
6. "Environmental Restoration and Waste Management - Five Year Plan, Fiscal Years 1993 - 1997," U.S. Department of Energy, Washington, DC, 1991, FYP DOE/S-0089P.

7. "Evaluation of Mid-to Long Term Basic Research for Environmental Restoration," Office of Energy Research, U.S. Department of Energy, Washington, DC, 1989, DOE/ER-419.
8. "DOE LDR Strategy Report for RMW," U.S. Department of Energy, Washington, DC, 1989, DOE/EH-91002927.
9. Portney, P., Report prepared for Resources for the Future, Washington, DC, cited in *Tox. Mat. News*, 1991, *18(45)*, 446.
10. Russell, M., "Hazardous Waste Remediation: The Task at Hand," Report for The Waste Management Research and Education Institute, The University of Tennessee, Knoxville, TN. Cited in *Tox. Mat. News*, 1991, *18(50)*, 493.
11. Cooke, S. M., *The Law of Hazardous Waste*; Mathew Bender and Co.: Inc., New York, NY, 1991; Vol. 1, pp. iv.
12. Rowe, Jr., W. D., "Superfund and Groundwater Remediation: Another Perspective," *Environ. Sci. Technol.*, 1991, *25(9)*, 370.
13. Travis, C.; Doty, C. "Commentary on Groundwater Remediation at Superfund Sites," *Environ. Sci. Technol.*, 1990, *24(10)*, 1464.
14. Wright, A. P.; Coates, H. A., "Legislative Initiatives for Stabilization/Solidification of Hazardous Wastes," *Toxic and Hazardous Waste Disposal*; Ann Arbor Science Publishers Inc.: Ann Arbor, MI, 1979; Vol. 2, Chapter 1.
15. Wagner, T. P., *The Complete Guide to The Hazardous Waste Regulations*; Second Edition; Van Nostrand Reinhold: New York, NY, 1991.
16. "Environmental Guidance Program Reference Books," Oak Ridge National Laboratory, Oak Ridge, TN, 1990, ORNL/M-1277.
17. *Environmental Law Handbook*; Eleventh edition; Government Institutes, Inc.: Rockville, MD, 1991.
18. *Environmental Statutes*; Eleventh edition; Government Institutes, Inc.: Rockville, MD, 1991.
19. "State Environmental Law Handbooks," These are state-specific handbooks on environmental law. Information on these can be obtained by calling (301) 251-9250.
20. *NEPA Deskbook*, Environmental Law Institute: Washington, DC, 1989.
21. EPA Register Notice, July 3, 1986. Interpreted by the DOE in 1987.
22. "DOE order 5820.2a," U.S. Department of Energy, Washington, DC, 1988.
23. Riley, R. G.; Zachara, J. M., "Nature of Chemical Contaminants on DOE Lands and Identification of Representative Contaminant Mixtures for Basic Subsurface Science Research," Office of Energy Research, U.S. Department of Energy, Washington, DC, 1992, DOE/ER-0547T.
24. R. E. Jacquish and R. W. Bryce, "DOE derived Concentration guides based on effective dose limit not to exceed 100 millirem/year." Derived from DOE Order 5480.1A: in "Hanford Site Environmental; Report for Calendar Year 1989" Pacific Northwest Laboratory, Richland, WA, 1990, PNL-7346.
25. Mackay, D. M.; Cherry, J. A., "Groundwater Contamination: Pump-and-Treat Remediation," *Environ. Sci. Technol.*, 1989, *23*, 630.

26. Mackay, D. M.; Roberts, P. V.; Cherry, J. A., "Transport of Organic Contaminents in Groundwater," *Environ. Sci. Technol.*, 1985, *19*, 384.
27. McCarty, P. L.; Reinhard, M.; Rittmann, B. E., "Trace Organics in Groundwater," *Environ. Sci. Technol.*, 1981, *15*, 41.
28. Mongan, T. R., "Priority Pollutant Criteria and the Clean Water Act," *Wat. Env. Tech*, 1991, December, *98*.
29. Perry, A. S.; Muszkat, L.; Perry, R. Y., "Pollution Hazards from Toxic Organic Chemicals," in "Toxic Organic Chemicals in Porous Media," Eds: Z. Gerstl, Y. Chen, U. Minglelgrin, B. Yaron, Springer-Verlag, Berlin, New York, 1989.
30. Heilshorn, E. D., "Removing VOCs from Contaminated Water," *Chem. Eng.*, 1991, *98*, 152.
31. Meyer, G., "Groundwater Contamination - No 'Quick Fix' in Sight," USGS yearbook; U.S. Geological Survey: Washington DC, 1980.
32. "Opportunities in Chemistry," Report of the Committee to Survey Opportunities in the Chemical Sciences, United States National Research Council, National Academy Press, Washington, DC, 1985.
33. "Frontiers in Chemical Engineering: Research Needs and Opportunities," Report of the Committee on Chemical Engineering Frontiers: Research Needs and Opportunities, United States National Research Council, National Academy Press, Washington, DC, 1988.
34. Ziegler, G. J., "Remediation Through Groundwater Recovery and Treatment," *Poll. Eng.*, 1989, *July*, 75.
35. "USEPA Treatability Manual: Volume III. Technologies for Control/Removal of Pollutants," Office of Research and Development, U.S. Environmental Protection Agency, Washington, DC, 1983, EPA-600/2-82/001a.
36. O'Neill, E. J., "Working to Increase the Use of Innovative Cleanup Technologies," *Wat. Env. Tech.*, 1991, *December*, 48.
37. Dworkin, D.; Cawley, M., "Aquifer Restoration: Chlorinated Organics Removal Considerations in Proven vs. Innovative Technology," *Env. Prog.*, 1988, *7*, 99.
38. Cheremisinoff, P. N., "Water and Wastewater Treatment - Fundamentals and Innovations," *Poll. Eng.*, 1989, *March*, 94.
39. Hauck, J.; Masoomian, S., "Alternative Technologies for Wastewater Treatment," *Poll. Eng.*, 1990, *May*, 81.
40. Eaton, D. L.; Smith, T. H.; Clements, T. L.; Hodge, V., "Issues in Radioactive Mixed Waste Compliance with RCRA: Some Examples from Ongoing Operations at the Idaho National Engineering Laboratory," Idaho National Engineering Laboratory, Idaho Falls, ID, 1990, EGG-M-89352.
41. McGlochlin, S. C.; Harder, R. V.; Jensen, R. T.; Pettis, S. A.; Roggenthen, D. K., "Evaluation of Prospective Hazardous Waste Treatment Technologies for Use in Processing Low-Level Mixed Wastes at Rocky Flats," EG&G Rocky Flats, Inc., Rocky Flats Plant, Golden, CO, 1990, RFP-4264.

42. "A Compendium of Technologies Used in the Treatment of Hazardous Waste," Center for Environmental Research Information, United States Environmental Protection Agency, Cincinnati, OH, 1987, EPA/625/8-87/014.
43. "Synopses of Federal Demonstrations of Innovative Site Remediation Technologies," Center for Environmental Research Information, United States Environmental Protection Agency, Cincinnati, OH, 1991, EPA/540/8-91/009.

RECEIVED June 29, 1992

GROUNDWATER AND SOIL DECONTAMINATION

Chapter 2

Removal of Inorganic Contaminants from Groundwater

Use of Supported Liquid Membranes

R. Chiarizia[1,3], E. P. Horwitz[1], and K. M. Hodgson[2]

[1]Chemistry Division, Argonne National Laboratory, 9700 South Cass Avenue, Argonne, IL 60439
[2]Westinghouse Hanford Company, P.O. Box 1970, Richland, WA 99352

This review paper summarizes the results of an investigation on the use of supported liquid membranes for the removal of uranium(VI) and some anionic contaminants (technetium(VII), chromium(VI) and nitrates) from the Hanford site groundwater. As a membrane carrier for U(VI), bis(2,4,4-trimethylpentyl)phosphinic acid was selected because of its high selectivity over calcium and magnesium. The water soluble complexing agent 1-hydroxyethane-1,1-diphosphonic acid was used as stripping agent. For the anionic contaminants the long-chain aliphatic amines Primene JM-T (primary), Amberlite LA-2 (secondary) and trilaurylamine (tertiary) were investigated as membrane carriers. Among these amines, Amberlite LA-2 proved to be the most effective carrier for the simultaneous removal of the investigated anion contaminants. A good long-term stability (at least one month) of the liquid membranes was obtained, especially in the uranium(VI) removal.

A supported liquid membrane (SLM) process has been considered, among other possible options, for the removal of contaminants from groundwater, because of the following advantages of SLM's over competing techniques (solvent extraction, ion exchange, polymeric membrane processes, etc.):

1. high concentration factors achieved through a high feed to strip volume ratio
2. low carrier inventory required
3. no phase separation problems
4. negligible organic phase entrainment in the feed and strip aqueous phases (although loss of organic phase due to solubility is still inevitable)
5. simplicity of operation of membrane modules.

These advantages, however, are balanced by typical drawbacks of SLM processes, such as the lack of a scrub stage, which makes more stringent the need of a high

---

[3]On leave from the Italian Nuclear and Alternative Energy Agency, ENEA, Rome, Italy

selectivity, and the lack of long-term stability, which allows for a practical application. With these considerations in mind, we have performed an investigation on the use of SLM's to remove selected contaminants from the Hanford site groundwater, as an application of the basic knowledge previously acquired at the Chemistry Division of Argonne National Laboratory (1-4). The detailed results of this investigation have been the subject of a number of publications (5-8). In the present paper we review the most important results and conclusions.

Groundwater

The detailed analysis of the Hanford groundwater has been reported in ref. (6). In Table I we report the typical concentrations of the species relevant for our investigation.

Table I. Concentration of Selected Contaminants in Hanford Groundwater

Contaminant	Concentration		MCL
	low	high	(maximum contaminant limit)
nitrate (ppm)	46	1,460	45
chromate (ppb)	51	437	50
^{99}Technetium (pico Ci/L)	906	29,100	900
Uranium (ppb)		8,590	10

SOURCE : Adapted from ref. (7).

To perform our SLM experiments, a synthetic groundwater solution (SGW), simulating the composition of the groundwater from a specific Hanford monitoring well, was prepared using the procedure reported in ref. (5). The composition of the SGW is reported in Table II. The pH of the SGW was adjusted to 2 with H_2SO_4 for reasons that will be discussed later. For distribution and/or permeation experiments, the SGW was spiked with U-233, or Tc-99, or made 10^{-3} $\underline{M}$ with Na_2CrO_4. From Table II it appears that, apart from sodium, the major cationic constitutents of SGW are calcium and magnesium. Any method devised to remove uranium from the solution, must therefore exhibit a very high selectivity over these two components. Similarly, a good selectivity for nitrates over sulfate-bisulfate species is required.

Table II. Composition of Synthetic Groundwater at pH 2

Constituent	Molarity
Calcium	0.012
Magnesium	0.0062
Sodium	0.017
Silicon	0.0009
Chloride	0.0016
Sulfate-bisulfate	0.017
Nitrate	0.030
Uranium	0.0004
Sum of Molarities	0.094

SOURCE: Reprinted with permission from ref. (5). Copyright 1990.

Membrane Supports

The liquid membrane supports were used both in flat-sheet and hollow-fiber configurations. In the flat-sheet membrane experiments Celgard or Accurel polypropylene membranes were used, with a thickness ranging from 25 to 100 microns, a pore size from 0.02 to 0.1 microns, and a porosity from 38 to 75%. The hollow-fibers were obtained from Enka. They were also made of propylene, with a porosity of 75%, a pore size of 0.1 microns, a wall thickness of 200 microns (I.D. = 0.6 mm, O.D. = 1 mm). The hollow-fibers were used to fabricate small laboratory scale modules, containing from 4 to 100 fibers about 10 cm long. For the tests with real groundwater discussed in the following, large size (2,600 fibers each 45.5 cm long) commercial Enka modules were used. The technique used to impregnate the supports with the carrier solution in n-dodecane has been described in refs. (5) and (6). All hollow-fiber modules were operated in recirculating mode, with feed and strip solutions flowed through the lumen and the shell side of the fibers, respectively, by calibrated peristaltic pumps. Other experimental details concerning the hydrodynamic conditions used in the flat-sheet and hollow-fiber experiments can be found in (5-8).

Uranium(IV) Removal from Synthetic Groundwater

The challenge of uranium(VI) removal from groundwater consists in finding a compromise between the two somewhat contradictory requirements of high selectivity for U(VI) over Ca(II) and Mg(II) and minimum adjustment of the feed composition. The latter requirement means that neutral and basic extractants (for example, mono- or bi-functional organophosphorus compounds and tertiary amines), showing a high affinity for U(VI), cannot be used as carriers, because they require high concentrations of anions such as nitrate for an effective uranium extraction. Organophosphorous acids, on the other hand, would extract from groundwater at neutral pH not only U(VI) but at least some significant amounts of all other cations with very little selectivity. A compromise solution, suggested by us in (5), was to add sulfuric acid to the groundwater lowering the pH to about 2, where very little Ca and Mg extraction takes place with organophosphorus acids. We thought that this low pH value would not only provide the required selectivity for U(VI) removal, but would also provide the hydrogen ions needed for the subsequent removal of the anionic contaminants in the form of acids by means of a basic carrier in a second membrane module. Of course, at the end of the process, the groundwater should undergo a neutralization step, before being again pumped under ground.

   Among some organophosphorous acids tested as U(VI) carriers from SGW at pH 2, bis(2,4,4-trimethylpentyl)phosphinic acid (H[DTMPeP]), contained in the commercial extractant Cyanex 272, was selected for its superior ability to reject calcium and magnesium. For example, with a 0.1 $\underline{M}$ solution of Cyanex 272 in n-dodecane, it was determined, from distribution experiments, that the selectivity for U(VI) over Ca(II), measured as the ratio of distribution ratios, was ~ 10^9. The next step was to find a suitable stripping agent capable of removing U(VI) from the SLM at the membrane-strip solution interface. The water soluble strong uranium(VI) complexing agent 1-hydroxyethane-1,1-diphosphonic acid, HEDPA, was found to be very effective. The distribution ratio of U(VI) between 0.1 $\underline{M}$ Cyanex 272 in n-dodecane and 0.1 $\underline{M}$ HEDPA in water was measured to be $6x10^{-3}$, that is, at least 3 orders of magnitude lower than with a 0.1 $\underline{M}$ solution of oxalic acid. Note here that the use of sodium carbonate, the traditional stripping agent for U(VI) in many solvent extraction studies, produced very short-lived liquid membranes, and, therefore, cannot be considered for the present application. The detailed description of the equilibria involved in the extraction of U(VI) by Cyanex 272 from SGW, and in its stripping by HEDPA, is reported in (*5*).

Uranium(VI) Permeation Studies. Figure 1 shows some typical results of permeation experiments, where the decrease of U(VI), Ca(II) and Fe(III) concentration in the feed is reported as function of time. The concentration data fall on straight lines described by the equation

$$\ln \frac{C}{Co} = -\frac{A}{V} P t \qquad (1)$$

where C and Co are the feed concentrations of transported species at time t and zero, respectively, A is the membrane area, V is the volume of the feed solution, and P the permeability coefficient (cm s^{-1}).

In the experiments of Figure 1, Fe(III) was also studied because this cation is ubiquitous and therefore its behavior is important even though it is not listed as a constituent in the Hanford site groundwater. Figure 1 shows that when 99% of U(VI) is removed from the SGW, after 2.0 hours, only 0.02% of the calcium follows the uranium. This corresponds to a membrane selectivity for U(VI) over Ca(II) equal to 1.6×10^4 (ratio of permeability coefficients). The data of Figure 1 were obtained with a membrane area equal to 9.8 cm^2 and a feed volume equal to 13 cm^3. For a much higher value of the A/V ratio, as usually provided by industrial hollow-fiber modules, the time required for the same level of uranium separation would be correspondingly shorter, but the relative contamination of uranium with calcium and iron, that depends on the selectivity, would be the same.

Figure 2 shows how the U(VI) permeability coefficients varies with the concentration of the membrane carrier (data obtained with flat-sheet supports). A striking feature of the data is the almost independence of P_U from the carrier concentration over about three orders of magnitude. A membrane initially containing 0.1 M Cyanex 272 in dodecane will continue to operate satisfactorily even when 99% of the carrier is lost due to solubility or other causes. The consequence of this result on the long-term membrane stability is evident. The continuous line of Figure 2 has been calculated with equation 2

$$P_U = \frac{D_U}{D_U \Delta_a + \Delta_o} \qquad (2)$$

where D_U = distribution ratio of U(VI) between feed and liquid membrane, $\Delta_a = d_a/D_a$ = thickness of aqueous diffusion layer/aqueous diffusion coefficient, cm-s^{-1}, and $\Delta_o = d_o/D_o$ = membrane thickness/membrane diffusion coefficient, cm-s^{-1}. How Equation 2 can be used to calculate the aqueous and organic diffusion coefficients of the U(VI) containing species is discussed in detail in (*5*).

To demonstrate that high concentration factors of U(VI) can be reached in practice with our SLM system, experiments were performed in which a 2 L solution of SGW was circulated in a module as the feed, while the strip solution (0.1 M HEDPA) had a volume of 45 mL. After six hours, uranium had been concentrated by a factor 34 in the strip solution. Much higher concentration factors (at least 10^3) can be achieved, however, by using the same strip solution over and over again. We have demonstrated in (*5*) that a 0.5 M HEDPA stripping solution, containing 0.2 M U(VI), is still very effective in stripping uranium from 0.1 M Cyanex 272 in n-dodecane.

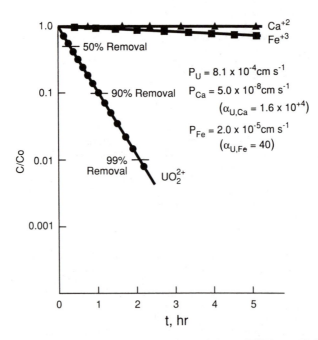

Figure 1. U(VI), Ca(II) and Fe(III) removal from SGW at pH 2. Liquid membrane = 0.1 M Cyanex 272 in n-dodecane; Strip = 0.1 M HEDPA; membrane area (hollow-fibers) = 9.8 cm^2; feed volume = 13 cm^3; feed linear velocity = 8.0 cm s^{-1}.

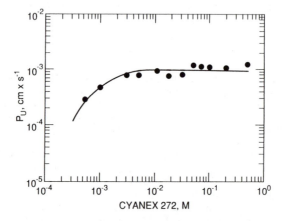

Figure 2. Permeability coefficient of U(VI) vs. carrier concentration. Feed = SGW at pH 2. Membrane: Cyanex 272 in n-dodecane on flat-sheet support; Strip = 0.1 M HEDPA. (Reproduced with permission from ref. (5). Copyright 1990 M. Dekker, Inc.)

Removal of Anionic Contaminants from Synthetic Groundwater

After passing through a SLM module, in which the uranium separation has taken place, the pH of the groundwater has not been significantly changed by the UO_2^{2+} - H^+ exchange with the phosphinic acid. As a consequence, the acidic pH of the groundwater can be exploited to remove nitrate, pertechnetate and chromate anions in the form of acids in a second SLM module, containing as carrier a basic molecule capable of reacting with these acids to form membrane soluble salts. After diffusing through the liquid membrane, these salts can be released at the strip side of the membrane, where an alkaline stripping solution (NaOH) ensures that the free carrier is regenerated.

Three commercially available long-chain aliphatic amines, Primene JM-T (primary), Amberlite LA-2 (secondary), and trilaurylamine (TLA, tertiary), were tested as membrane carriers for nitrate, pertechnetate and chromate anions. Long-chain aliphatic amines, dissolved in an organic diluents, are known to extract acids according to the reaction

$$n\,H^+ + A^{n-} + n\,\overline{B} \overset{K}{\rightleftharpoons} \overline{(BH)_nA} \rightleftharpoons \overline{\text{aggregates}} \qquad (3)$$

where H_nA is a generic acid in the aqueous solution, B is the amine, and the bar represents organic phase species. K is the equilibrium constants that can be taken as a measure of the affinity of the amine for the acid. Table III summarizes the physico-chemical properties of the three amines investigated, of relevance for the choice of the membrane carrier for our specific application.

Table III. Properties of Primene JM-T (I), Amberlite LA-2(II) and TLA(III)

Chemical affinity	for HNO_3	:	$K_I > K_{II} > K_{III}$
	for $HTcO_4$	:	$K_{III} \geq K_{II} > K_I$
	for H_2CrO_4	:	$K_{III} \geq K_{II} > K_I$
Solubility in water		:	$S_I > S_{II} > S_{III}$
Interfacial pressure		:	$\Pi_I > \Pi_{II} > \Pi_{III}$

The detailed determination of the equilibrium constants and of the interfacial behavior of the three amines shown in Table III are reported in refs. (7-8). From the data of Table III it appears that the primary amine would not be a good choice as a carrier because, although a better extractant for HNO_3, it is a relatively poor extractant for the other two acids. Also, its higher solubility in water and its greater lowering of the interfacial tension (higher tendency to emulsion formation) are an indication that SLM containing Primene JM-T would be more unstable. The tertiary amine, on the other hand, showing the lowest solubility in water and the best interfacial behavior, exhibits the lowest affinity for HNO_3. It seems, therefore, that the best compromise among the properties listed in Table III is the choice of Amberlite LA-2 as the membrane carrier.

Anionic Species Permeation Studies. The dependence of the HNO_3 permeability coefficient on the concentration of the three amines in n-dodecane (flat-sheet membrane experiments) is reported in Figure 3. It appears from the data that,

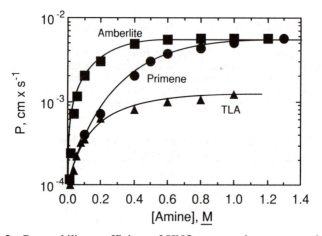

Figure 3. Permeability coefficient of HNO_3 vs. carrier concentration in n-dodecane. Feed = 10^{-2} $\underline{M}$ HNO_3; Membrane = flat-sheet support; Strip = 0.1 $\underline{M}$ NaOH. (Reproduced with permission from ref. (7). Copyright 1991 Elsevier Science Publishers B.V.)

with the primary and secondary amine, the same limiting P value is reached. Amberlite LA-2, however, reaches the limiting value at a much lower concentration and, therefore, following the same reasoning as for the Cyanex 272 case in Figure 2, is a better carrier for nitric acid. The behavior of Primene JM-T, which has a higher equilibrium constant for HNO_3 extraction (see Table III), may be due to its higher solubility in water, or simply may reflect the extreme complexity of the aggregation equilibria in the organic phase. The HNO_3 permeability coefficient with TLA as carrier is always much lower than for the other two amines, except for very low carrier concentrations. This may indicate that a local precipitation of the nitrate-TLA salt takes place in the pores of the membrane, reducing the speed of permeation.

In the groundwater acidified at pH 2 with sulfuric acid, sulfate and bisulfate anions are present. By using an amine as the carrier in a SLM system, it is important to know what fraction of the total H^+ is transported by the liquid membrane as HNO_3, because we are interested in removing nitrates, not sulfates, from the groundwater. For this purpose a number of experiments were performed where a pH electrode and a nitrate electrode were used to follow the decrease of acidity and of nitrates in the acidified SGW used as the feed. A detailed discussion of the results is reported in (7). Here it is important only to mention that the removal of nitrates, with all three amines, followed quite closely the removal of total acid. With 0.6 M Amberlite LA-2 as carrier, for example, when 90% of H^+ was removed, about 75% of the removable nitrate ions had left the feed. This result allowed us to conclude that H^+ is removed from the SGW mainly as nitric acid and that amine based SLMs are effective in removing nitrates from SGW even in the presence of large quantities of sulfate-bisulfate anions.

Membrane experiments showed that the efficiency of the three amines as carriers for Tc(VII) parallels the sequence of equilibrium constants reported in Table III. That is, secondary and tertiary amines are better at removing Tc(VII) than primary amines.

An unexpected result was found investigating the Cr(VI) removal by the three amines. Contrary to the sequence of Table III, the use of TLA as carrier led to a very low value of the Cr(VI) permeability coefficient, even lower than with Primene JM-T. This result is probably a further indication of the poor solubility of TLA salts (in this case chromate) in n-dodecane.

In conclusion, the permeation behavior of the anionic contaminants under study through SLMs containing either one of the amines investigated, confirms that the carrier of choice for the simultaneous removal of nitrates, Tc(VII) and Cr(VI) is the secondary amine Amberlite LA-2.

Tests with Real Groundwater

Some tests with 50 gallons samples of groundwater from a specific monitoring well were performed at Hanford using two 2.2 m^2 internal surface area commercial hollow-fiber modules in series, containing a 0.1 M Cyanex 272 and a 0.1 M Primene JM-T solution in n-dodecane, respectively, as liquid membranes. The stripping solutions were 4 gallons of 0.1 M HEDPA for the first module, and 4 gallons of 0.1 M NaOH for the second one. Other experimental details are reported in (6). The result of a typical test are summarized in Table IV.

The results show that the two modules were very effective in reducing the U(VI) and Tc(VII) concentrations by about three and two orders of magnitude, respectively. The limited success with NO_3^- is due to the fact that these tests were performed before the final choice of the best carrier for nitrates was made. The use of a 0.2 M Amberlite LA-II solution as liquid membrane in the second module would have led to much better results for the removal of nitrates.

Table IV. Results of Tests with Real Groundwater

Time (hours)	First Module Feed U(VI)(ppb)	Second Module Feed NO₃ (ppm)	Feed Tc(VII)(pCi/L)
0	3,460	38.5	786
4	257	37	361
8	30	31	139
12	7.3	24.7	51
16	2.1	20.9	18
20	1.4	18.1	9
24	—	15.3	8
36	—	10.2	4
48	—	5.2	2

Flowrates: 1.5 gal/min, shell side, feed; 1.0 gal/min, lumen side, strip.

Liquid Membrane Stability

To test the ability of our liquid membrane system to continuously operate at high efficiency, stability tests were performed. They are described in detail in ref. (6) for the uranium removal from groundwater and in ref. (8) for the removal of the anionic contaminants. Some results are shown in Figure 4, as uranium permeability vs. time (a constant permeability was the criterion for stability), for two modules that were operated without interruption (except nights and weekends) for very long times. An excellent stability, that is constant uranium permeability, was shown by the module with reservoir (it contained a small reservoir of carrier solution ensuring a continuous reimpregnation of the membrane pores). The stability test with this module was interrupted after six months because of the deterioration of the reservoir seal. However, it worked long enough to demonstrate that a properly designed self-reimpregnating module can operate for a practically unlimited time. The conventional module (without reservoir) was periodically reimpregnated with the carrier solution, when the initial permeability of uranium had declined by about 50%. The procedure was repeated seven times over a time span of almost 1.5 years. The results reported in Figure 4 show that periodic reimpregnations of the hollow-fibers are also a viable technique to have modules operating efficiently for long times. It is interesting to note that the reimpregnation procedure, described in detail in (6), while not affecting the stability of the membrane, had a postive effect on the module performance, measured by the uranium permeability, which improved substantially and progressively after each of the first three reimpregnations. Different strip solutions were used after the last two reimpregnations. In one case a 1 M (instead, of 0.1 M) HEDPA solution was used to see if a much higher osmotic pressure difference between feed and strip solution would affect the stability, in the other case a different stripping agent, a derivative of HEDPA, was tested.

Stability tests were also performed with liquid membranes containing each of the three amines investigated as carrier for anions. In the experiments involving liquid membranes adsorbed on flat-sheet supports having a thickness of only 25 microns, the order of stability tertiary > secondary > primary was measured. This is the reverse order of amine solubility in water and of the interfacial lowering at a water-dodecane interface. Both factors, solubility and interfacial tension, seem to be operative in determining the liquid membrane stability, together with the other usual factors, such as support materials, pore size, osmotic pressure gradient and flow rate, which have been kept constant in this work. An impressively high stability was measured in the experiments with flat-sheet membranes, when TLA was the carrier. We think that the high stability of the TLA-dodecane liquid

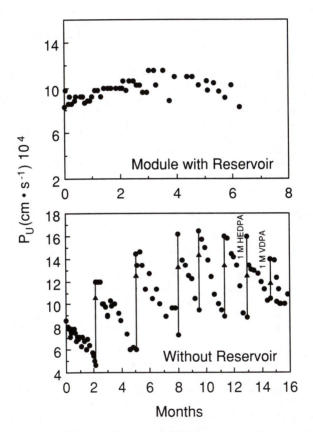

Figure 4.   Permeability coefficient of U(VI) vs. time for a module with reservoir and a conventional module. Feed = SGW at pH 2, 10 mL, replaced daily; Membrane = 0.1 $\underline{M}$ Cyanex 272 in n-dodecane on hollow-fibers; Strip = 0.1 $\underline{M}$ HEDPA, 16 mL, replaced monthly. (Reproduced with permission from ref. (*6*). Copyright 1990 M. Dekker, Inc.)

membrane is due to the low solubility of the amine salts in the membrane diluent. The formation of a solid or gelatinous phase in the membrane pores, while having a detrimental effect on the permeation speed, may have a positive effect on the membrane life, by preventing the formation of emulsion with the aqueous phase and by acting as a barrier against water bridging in the membrane pores.

Stability tests were also performed using modules containing hollow-fibers with a much higher wall thickness (200 microns). Because of the much larger inventory of organic phase contained in the pores of the fibers, a complete failure of the membrane over a time span of about 1 month was not observed for any of the three amines, although the initial flux of HNO_3 through the membranes was significantly reduced, especially with the primary amine. Also from a stability standpoint, however, the secondary amine Amberlite LA-2 exhibited the best performance, loosing only 40% of the initial HNO_3 flux after 40 days of continuous operation.

In conclusion, the stability tests performed with hollow-fiber modules containing as liquid membranes n-dodecane solutions of either Cyanex 272 (carrier for uranium(VI)) or Amberlite LA-2 (carrier for anionic contaminants) have shown that a relatively long membrane life, of the order of at least one month, can be expected in the processing of contaminated groundwater. After one month, or more, of continuous operation, the modules can be easily reimpregnated without loosing their separation efficiency. Alternatively, self-reimpregnating modules seem capable of operating without problems for much longer times.

Conclusions

The experiments reported in ref. (5-8) have shown that a few options are available for selecting a supported liquid membrane process for the removal of U(VI), Cr(VI), Tc(VII) and nitrate ions from acidified groundwater. The first option involves the use of two modules in series: the first one, containing the extractant Cyanex 272 as carrier, removes uranium, while the second one, containing the long-chain aliphatic secondary amine Amberlite LA-2 as carrier, removes the anionic contaminants. We have demonstrated that this system is highly effective in achieving the desired decontamination and that membrane lives of at least one month of continuous operation can be expected (much longer membrane lives can be obtained using the somewhat more complicated self-reimpregnating modules).

An alternative option could be the use of the tertiary amine TLA as the carrier in the second module, because of the higher stability of TLA-based liquid membranes. In this case the benefit of less frequent module reimpregnations would compensate for the lower effectiveness of TLA in removing nitrates. The TLA alternative appears especially attractive if used in a combined process where the nitrate removal from groundwater is achieved mainly by other means (for example, biological).

A third alternative, which probably deserves more investigation, is the use of a single membrane module, instead of two in series, containing the primary amine Primene JM-T as carrier. This compound has the unique property of removing from groundwater acidified with sulfuric acid not only the anionic contaminants, but also U(VI) in the form of anionic sulfato-complexes (7). In this way all the unwanted contaminants would be removed in the same module, making the process simpler. The advantage of using one single module, however, would be counteracted by the need for more frequent reimpregnations, because of the shorter lifetime of the Primene JM-T based membrane, unless a self-reimpregnating module is used.

We think that all three options discussed above can be realized in practice. The choice among them should be based mainly on economic and engineering considerations.

Acknowledgments. The first two authors wish to express their gratitude to Westinghouse Hanford Co. for the financial support provided.

Literature Cited

1. Danesi, P. R.; *Sep. Sci. Technol.*, **1984-85**, *19(11-12)*, 857.
2. Danesi, P. R.; Chiarizia, R.; Rickert, P. G.; Horwitz, E. P.; *Solvent Extr. Ion Exch.*, **1985**, *3(1-2)*, 111.
3. Chiarizia, R.; Danesi, P. R.; *Sep. Sci. Technol.*, **1987**, *22(2-3)*, 641.
4. Danesi, P. R.; Reichley-Yinger, L.; Rickert, P. G.; *J. Membr. Sci.*, **1987**, *31*, 117.
5. Chiarizia, R.; Horwitz, E. P.; *Solvent Extr. Ion Exch.*, **1990**, *8(1)*, 65.
6. Chiarizia, R.; Horwitz, E. P.; Rickert, P. G.; Hodgson, K. M.; *Sep. Sci. Technol.*, **1990**, *25(13-15)*, 1571.
7. Chiarizia, R.; *J. Memb. Sci.*, **1991**, *55*, 39.
8. Chiarizia, R.; *J. Memb. Sci.*, **1991**, *55*, 65.

RECEIVED December 26, 1991

Chapter 3

Removal of Plutonium from Low-Level Process Wastewaters by Adsorption

G. S. Barney, K. J. Lueck, and J. W. Green

Westinghouse Hanford Company, Richland, WA 99352

Plutonium removal from low-level wastewater effluents at the Hanford Site was faster and more complete using a bone char adsorbent than for other commercially-available adsorbents tested. Equilibrium distribution coefficients (Kd values) were high (8,000 to 31,000 mL/g) for plutonium adsorption on bone char over the range of pH values expected in the wastewaters (5 to 9). Plutonium decontamination factors for wastewater flowing through columns of bone char ranged from 400 to 3,000 depending on the column flow rate. Other adsorbents tested included cation and anion exchange resins, activated alumina, and metal removal agents. Adsorption on the cation exchange resins was rapid, but a small fraction (~ 2%) of the plutonium was inert toward these resins. Anion exchange resins and alumina were effective adsorbents only at high pH levels.

Wastewater containing small concentrations of plutonium is generated during processing operations at the Plutonium Finishing Plant (PFP) on the Hanford Site in Washington State. This wastewater consists mainly of cooling water and is generated at a rate of about 6 to 11 million liters per month. A wastewater treatment facility is planned that will remove the plutonium to specified, low concentrations prior to discharge of the wastewater to the environment. The plutonium will be adsorbed on ion exchange resins or other types of adsorbents by passing the wastewater through columns containing beds of the adsorbent. The objective of the present study was to evaluate potential adsorbents for use in the treatment facility. The adsorbent must have a high affinity for plutonium species dissolved or suspended in the wastewater over the range of pH values expected, adsorption must be rapid, and the particle size of the adsorbent must be large enough to prevent a large pressure drop across the columns.

Laboratory and engineering studies have been performed earlier on a number of potential adsorbents. Roberts and Herald (1) tested a number of cation and anion resins and bone char for removal of plutonium from aqueous low-level processing wastes at Mound Laboratory in Miamisburg, Ohio. They found that bone char gave the highest decontamination factors at pH values of 3 and 7 in large-scale engineering column tests. Macroporous anion and cation resins also adsorbed plutonium well in this pH range. Adsorption of plutonium "polymer" on bone char from buffered

0097–6156/92/0509–0034$06.00/0

aqueous solutions was examined by Silver and Koenst (2) in laboratory tests and by Blane and Murphy (3) using pilot plant scale columns. The laboratory tests showed that plutonium adsorption is pH dependent and that adsorption on bone char is substantial over a wide pH range. Smith et al. (4) studied plutonium adsorption on a modified peat, zeolite (Type 4A), and on coconut carbon. The zeolite and peat were the most effective adsorbents over the pH range 4 to 10 as shown by column flow-through adsorption measurements. A macroporous weak base anion exchange resin (Amberlite XE270) was found to strongly adsorb Pu(IV) "polymer" from an aqueous solution with pH 4 to 10 (5). Schulz et al. (6) reported studies of adsorption of plutonium and americium on bone char and on a synthetic sodium titanate adsorbent. They found that both of these adsorbents removed alpha activity from near-neutral and caustic wastes to low levels.

Ten different adsorbents were chosen for the present evaluation. They include four types of adsorbents: (1) inorganic activated alumina and bone char adsorbents, (2) macroporous cation exchange resins, (3) macroporous anion exchange resins, and (4) chelating heavy metal removal agents. Equilibrium adsorption of Pu(IV) from actual PFP wastewater spiked with a ^{238}Pu tracer was measured for each of the adsorbents. The plutonium tracer was required because the activity of plutonium in the wastewater was too low to be measured conveniently. Those adsorbents that adsorbed plutonium strongly were then tested in column flow-through experiments and the rate of plutonium adsorption was measured.

Experimental

Materials. Actual PFP wastewater (taken from Manhole 9 on 8/24/88) was used in all adsorption experiments. The composition of this wastewater is given in Table I. Only the major components are shown. The pH of the wastewater was determined to be 8.3. The composition occasionally varies from that of the wastewater composition presented

Table I. PFP Wastewater Composition

Non-Radioactive Composition		Radioactive Composition	
Component	Concentration ($\mu g/L$)	Component	Concentration ($\mu Ci/L$)
Ca	18000	^{238}Pu	2×10^{-5}
Mg	4100	^{239}Pu	1×10^{-4}
Na	2400	^{241}Pu	9×10^{-4}
K	1300	^{241}Am	4×10^{-4}
Sr	226	^{90}Sr	2×10^{-5}
Al	190	^{137}Cs	4×10^{-5}
SO_4	1500		
Cl	3500		
NO_3	935		
F	114		
Chloroform	27		
Acetone	330		

in the table and depends on operations in the PFP facility. Analyses over a several year period show that it is essentially a dilute calcium sulfate solution with a pH ranging from 5 to 9. Total alpha activity varies from about 10^{-5} to 10^{-2} $\mu Ci/L$. The volume of wastewater generated varies from about 6 to 11 million L/month.

The alpha activity of plutonium in the wastewater sample was too low to be measured conveniently. A ^{238}Pu spike solution (with 95% ^{238}Pu activity and 5% ^{241}Pu activity) was added to the wastewater as a tracer for plutonium. The ^{238}Pu spike solution was chemically treated to assure that all the Pu was in the (IV) oxidation state before adding it to the wastewater. This was done by first reducing all the plutonium to Pu(III) with hydrogen peroxide (in 1.0 $\underline{M}$ HNO$_3$). The Pu(III) was then oxidized to Pu(IV) by adding sodium nitrite. The plutonium spike was assumed to react with the wastewater to form a plutonium species distribution similar to that of the original wastewater. Nitsche (7) has shown that Pu(IV) spikes added to groundwater react in a relatively short time to form mainly Pu(V) and Pu(VI) species.

The characteristics of adsorbents used in this study are summarized in Table II. There are four general types of adsorbents: (1) chelating heavy metal removal agents, (2) cation exchange resins, (3) anion exchange resins, and (4) activated alumina and carbon adsorbents. Some of these materials may not be practical for use in long columns because of their small granule size (e.g., Alumina-C and MLM-300).

Procedures. Procedures were developed for measuring plutonium distribution coefficients for batch equilibrium adsorption measurements and for measuring plutonium breakthrough concentrations for column flow-through adsorption experiments.

Equilibrium Distribution Measurements. Portions of the wastewater sample were spiked with two levels of ^{238}Pu activities (approximately 1 and 10 μCi/L). Total plutonium concentrations in the spiked solutions were about 40 to 400 times those of the original groundwater in terms of μg/L. Separate portions of the spiked wastewaters were adjusted to pH values of 5, 6, 7, and 8 using dilute NaOH or HNO$_3$ solutions as required. Each of the resulting solutions was filtered through 0.45 μm pore-size Millipore filters to remove any plutonium-containing solids and the plutonium concentrations in these filtered solutions were measured using a liquid scintillation analyzer. The pH of water in contact with each adsorbent was adjusted to 5, 6, 7, or 8 so that the adsorbent would not change the desired pH of the final equilibration with ^{238}Pu-spiked wastewater. The adsorbent-water mixture was filtered. Two-gram samples of each adsorbent (wet) adjusted to the proper pH were equilibrated with 10.0 ml of the plutonium-spiked wastewater at the proper pH and Pu concentration. The batch equilibrations were performed in triplicate over a period of 24 hours in 20 ml polypropylene screw-cap vials. The vials were then centrifuged (or filtered in the case of bone char) to separate the phases and a 2.00 ml sample of the solution was analyzed for plutonium concentration. The remaining solution was used to measure the final pH of the mixture.

Plutonium distribution coefficients for each adsorbent at four pH levels and two plutonium concentration levels were calculated as follows:

$$Kd \; (mL/g) = \frac{Pu \; activity/g \; of \; wet \; adsorbent}{Pu \; activity/mL \; of \; solution}$$

Average values of Kd for the metal removal agents, cation exchange resins, activated alumina and carbon adsorbents, and anion exchange resins are presented in the Results and Discussion section.

Flow-Through Column Measurements. All apparatus in contact with the wastewater solution was made of plastic (either Teflon or polypropylene) to avoid adsorption on glass walls. A Teflon cylinder with bed dimensions of 56 mm long x 25

Table II. Characteristics of the Adsorbents Studied

Adsorbent	Type	Chemical Form	Granule Size	Capacity	Useable pH Range
Bioclaim-MRA (Advanced Mineral Technologies)	Cation exchanger/complexant for heavy metal removal	Na$^+$	~ 0.1 mm	10% of Weight	4 - 10
Remsorb-CP (Pacific Nuclear Systems)	Chelating adsorbent for heavy metals	Na$^+$	Fine	–	–
Amberlite IRC-718 (Rhom and Haas)	Chelating cation exchanger for heavy metals	Na$^+$	16 - 50 mesh	1.0 meq/ml	2 - 10
Duolite GT-73 (Rohm and Haas)	Cation exchanger for Hg, Ag, Cu, Pb, and Cd removal	H$^+$	0.3 - 1.2 mm	1.2 meq/ml	1 - 13
Alumina-C (Universal Adsorbents)	Al$_2$O$_3$	Na$^+$	5 - 200 μm	–	Wide
MLM-300 (Universal Adsorbents)	Activated alumina adsorbent	Na$^+$	Fine	–	Wide
Bone Char (Stauffer Chemical)	Activated carbon/calcium phosphate adsorbent	Ca^{2+}	8 - 28 mesh	–	Wide
Lewatit MP-500-FK (Mobay Chemical)	Anion exchanger (macroporous, strong base)	Cl$^-$	40 - 50 mesh	–	Wide
Dowex 21K (Dow Chemical)	Anion exchanger (strong base, trimethylamine)	Cl$^-$	20 - 50 mesh	1.3 meq/ml	Wide
Dowex M-41 (Dow Chemical)	Anion exchanger (macroporous, strong base, trimethylamine)	Cl$^-$	20 - 50 mesh	1.0 meq/ml	Wide

mm diameter) fitted with tubing connectors and polypropylene mesh filter material was used as a laboratory column to contain the adsorbent during flow-through experiments. The adsorbent was pre-equilibrated with water at the desired pH and then slurried into the cylinder to prevent entrapment of air in the adsorbent bed. The pore volume of the adsorbent bed was determined by subtracting the weight of the empty cylinder plus the weight of the adsorbent added to the cylinder from the cylinder plus adsorbent saturated with distilled water.

Several liters of the wastewater to be used as feed to the columns were spiked with ^{238}Pu to a concentration of about 4 µCi/L. The pH was re-adjusted to the original 8.3 using dilute NaOH. This solution was pumped with an Eldex Model E-120-S precision metering pump through a 1.0 µm pore size Teflon Millipore filter and then to the bottom of a column prepared as in the previous paragraph. The effluent from the top of the column was collected in plastic tubes with an HBI Model LC 200 fraction collector. Each tube collected one pore volume of effluent. Six flow rates were used to determine the rates of adsorption: 0.5, 1, 2, 3, 4, and 5 ml/min. At least five consecutive pore volumes were collected for each flow rate to ensure steady state concentrations of plutonium in the effluent. The effluent concentration, C, was compared with the influent concentration, Co.

Analyses. Liquid scintillation counting was used to measure the ^{238}Pu activity in all samples. Aqueous samples (1.00 or 2.00 ml) were added to 18 ml of Insta-Gel scintillation cocktail (Packard Instrument Co.) in glass scintillation vials and thoroughly mixed. A Packard Tri-Carb Model 1550 liquid scintillation analyzer was used to count all the samples.

Results and Discussion

Plutonium Distribution Coefficients. Equilibrium distribution coefficients measured for adsorption of plutonium on most of the adsorbents were significantly affected by solution pH. Solution pH has a large effect on the identity of plutonium chemical species in solution and on the chemical form of surface adsorption sites on the adsorbents. These factors will, of course, affect adsorption reactions of plutonium with adsorbent surfaces. Dissolved plutonium in these wastewaters should exist mainly as hydroxyl and carbonate complexes (8). Although carbonate concentration was not measured, it can be presumed that the wastewater contains carbonate from carbon dioxide absorption from the air and carbonate originally present in the well water. At the higher pH range studied (pH 8 to 9), plutonium will exist in solution mainly as anionic complexes such as $PuO_2(CO_3)_3^{5-}$ or $Pu(OH)_5^-$. Cations and uncharged species predominate at lower pH values. Polymeric plutonium hydroxide may be present in the wastewater, but its formation in significant quantities seems unlikely because of the low plutonium concentrations (10^{-11} to 10^{-9} M) in the wastewater (9). These concentrations are below the solubility of plutonium hydroxide polymer which has been measured at greater than 10^{-7} M at pH 8.5 to 11 (10). Additional evidence against the existence of polymer is the fact that most of the plutonium in the wastewater is adsorbed by cation exchange resins. Ockenden and Welch (11) have shown that the polymer is not adsorbed by cation resins.

Plutonium adsorption distribution coefficients for two alumina adsorbents, Alumina-C and MLM-300, are presented in Figure 1. Error bars in all figures show the standard deviations of three measurements for each data point. Values of Kd increase greatly as the solution pH increases. The slope for both plots is approximately one. This suggests that one H^+ was released for each plutonium ion adsorbed (12). The source of the H^+ is uncertain. It could be released from surface OH groups or from the hydrolyzed plutonium species during adsorption. The increased adsorption at higher pH is normal behavior for adsorption of hydrolyzed metal ions on metal oxide surfaces (12). Since distribution coefficients for the two plutonium concentration levels are essentially the same, the adsorption isotherm must be nearly linear in this concentration range. The usefulness of the alumina adsorbents for plutonium removal from wastewater is limited because of the relatively weak adsorption at low pH.

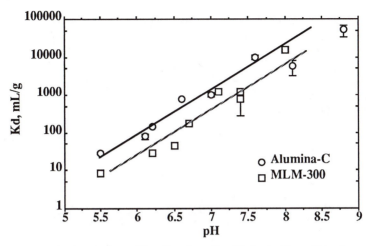

Figure 1. Plutonium Distribution Coefficients For Alumina Adsorbents as a Function of pH.

The distribution coefficients are very high (10,000 to 50,000 mL/g) for wastewater with a pH ranging from 8 to 10. However, much of the wastewater will have a pH less than 6. The small particle size of these particular adsorbents would also limit wastewater flow through large wastewater treatment columns.

Plutonium distribution coefficients for adsorption on bone char are quite high, but reach a maximum at about pH 8 (Figure 2). The major component of bone char, $Ca_3(PO_3)_2$, buffered the wastewater solution so that it was not possible to achieve a pH of less than 7. Interestingly, the pH of maximum adsorption is the same as the minimum solubility of beta-tricalcium phosphate (*13*), the major component of bone char. This implies that plutonium adsorption may be diminished by phosphate dissolved in the wastewater. As with the alumina adsorbents, the Kd values for the two plutonium concentrations are not significantly different and thus, the isotherm must be linear. The particle size of bone char was large enough so that reasonable flow rates through large columns should be easily achieved.

Figures 3 and 4 show Kd values observed for two macroporous cation exchange resins, Duolite GT-73 and Amberlite IRC-718. The adsorption of plutonium on both resins was relatively weak compared with other adsorbents, especially at the lower initial plutonium concentrations. At initial concentrations of about 1 μCi/L, Kd values for Duolite GT-73 and Amberlite IRC-718 were approximately 200 and 76 ml/g, respectively. Surprisingly, no dependence on pH is apparent at the lower initial plutonium concentrations. More cationic plutonium species should be present at lower pH values and would be more strongly adsorbed on cation exchange resin. This effect is only observed for Duolite GT-73 at the higher initial plutonium concentration, 10 μCi/L. The absence of a pH effect at the lower concentration can be explained by the existence of a small concentration of a stable plutonium species in solution that does not interact with cation resins. Additional evidence supporting this explanation is the fact that about 1.3 to 2.8% of the initial plutonium is not adsorbed on the cation resins in both the batch equilibrium adsorption measurements and the column adsorption experiments discussed later. This small fraction of plutonium is not adsorbed over a

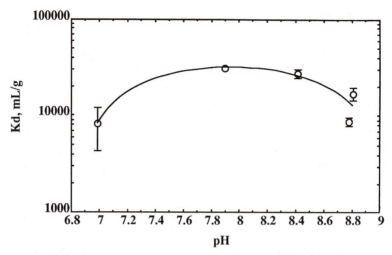

Figure 2. Plutonium Distribution Coefficients for the Bone Char
Adsorbent as a Function of pH.

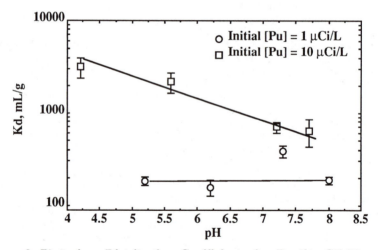

Figure 3. Plutonium Distribution Coefficients for Duolite GT-73 as a
Function of pH.

wide range of solution pH values and reaction times and is unreactive toward the cation
resins. This "inert" plutonium could be a strongly complexed neutral or anionic species
or perhaps a small amount of Pu(IV) polymer. Since the cation resins do not react with
and remove all significant plutonium species from the wastewater, their effectiveness in
wastewater treatment is limited.

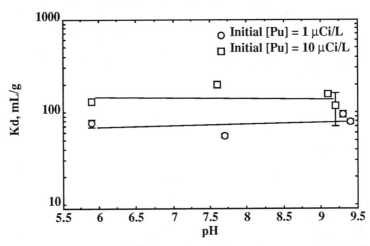

Figure 4. Plutonium Distribution Coefficients for Amberlite IRC-718 as a Function of pH.

Figure 5 shows plutonium adsorption Kd values for three anion exchange resins over a pH range of 4 to 8. Plutonium adsorption on each of the three resins, Lewatit MP-500-FK, Dowex 21K, and Dowex M-41, increased significantly as the pH increased. This was expected since anionic hydoxyl and carbonate complexes are more stable at higher pH. Since these resins are not effective plutonium adsorbents below a pH of 7 or 8, they are not useful in wastewater treatment (at least when used alone).

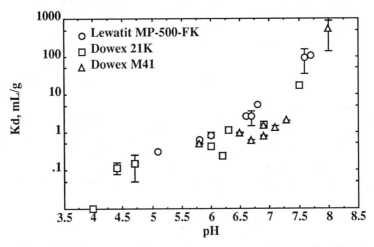

Figure 5. Pu(IV) Distribution Coefficients for Three Anion Exchange Resins as a Function of pH.

The final class of adsorbents studied were the chelating heavy metal removal agents. These materials are chemically modified biological products that strongly bind heavy metals by ion exchange and chelating mechanisms. The Bioclaim-MRA adsorbent is derived from the cell walls of bacteria and Remsorb-CP from shellfish. Figure 6 shows plutonium adsorption Kd values for these materials. The Kd values for Bioclaim-MRA were very high (approximately 4000 to 7000 ml/g) and showed no dependence on pH.

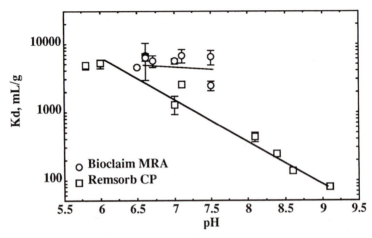

Figure 6. Pu(IV) Distribution Coefficients for Chelating Heavy Metal Removal Agents as a Function of pH.

The pH range achieved was very narrow (6.5 to 7.5) because Bioclaim-MRA was a strong buffer. Distribution coefficients for plutonium adsorption on Remsorb-CP were somewhat lower and decreased dramatically as the pH increased to 9 as might be expected for a cation exchanger. There was no significant difference in Kd values for the two
different initial plutonium concentrations. This suggests that the adsorption isotherms for these adsorbents are linear over the concentration range studied. The particle size of Bioclaim-MRA is large enough for use in water treatment columns; however, a black algal or bacterial growth was observed in some experiments after a period of one week. This growth caused some of the granules to stick together, forming strings. The growth may not occur as rapidly in columns containing flowing wastewater, but it potentially could plug the columns. The strong plutonium adsorption on Bioclaim-MRA and lack of influence of solution pH make this adsorbent attractive for wastewater treatment. Relatively low plutonium Kd values at expected wastewater pH values and a small particle size limit the usefulness of the Remsorb-CP adsorbent for wastewater treatment.

Column Adsorption Measurements

Column flow-through measurements of plutonium adsorption from wastewater allow determination of the relative rates of adsorption by the different adsorbents. Plutonium concentrations in the column effluents were measured for six different flow rates using four of the most promising adsorbents: Bioclaim-MRA, Duolite GT-73, Amberlite IRC-

718, and bone char. The results of these measurements are given in Figure 7 which is a plot of C/Co versus flow rate where Co is the influent plutonium concentration and C is the steady-state effluent concentration. Although Bioclaim-MRA had relatively high adsorption Kd values, the rate of adsorption was slow. Bone char adsorbed plutonium rapidly and more completely than the other adsorbents. Adsorption on Amberlite IRC-718 was almost independent of flow rate. As discussed above, this is due to a small amount of some unreactive plutonium species that passes through the column without

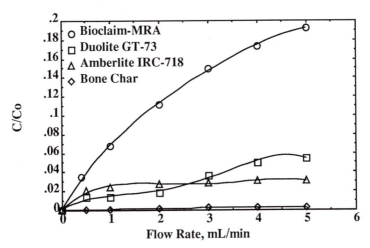

Figure 7. Plutonium Concentrations in Adsorbent Column Effluents for Various Flow Rates

interacting with the resin. The majority of the reactive plutonium is adsorbed rapidly since none of it appears in the column effluent even at the highest flow rate. Plutonium effluent concentrations from a Duolite GT-73 column were similar to those for Amberlite IRC-718. Adsorption was somewhat slower, however, since the concentration increased at the higher flow rates. Plutonium decontamination factors for laboratory columns over the range of flow rates used in these experiments are given in Table III.

Table III. Plutonium Decontamination Factors (DF's) for Column Tests

Adsorbent	pH	DF
Bone Char	9.1	377 - 3,125
Duolite GT-73	6.8	18 - 77
Amberlite IRC-718	9.5	33 - 50
Bioclaim-MRA	10.0	5 - 29

Bone char appears to be the most effective adsorbent for removal of plutonium from the wastewaters because of high distribution coefficients for plutonium adsorption and rapid adsorption. The capacity of bone char for plutonium absorption can be calculated from a measured Kd value, assuming a reversible equilibrium adsorption reaction and a linear adsorption isotherm. If the capacity is defined as the point where the plutonium concentrations in the column feed and effluent are equal, the maximum volume of wastewater that can be treated (V_{max}) is (*14*)

$$V_{max} = Kd \text{ x (Weight of Bone Char).}$$

Assuming a Kd value of 30,000 mL/g, one kg of bone char will treat 30,000 L of wastewater. This capacity was approached during tests with a small column of bone char containing 2.3 g of this adsorbent with bed dimensions of 8 mm x 36 mm long. Approximately 30,723 mL of wastewater spiked with ^{238}Pu was passed through the column over a period of several months at a flow rate of 0.5 mL/min. Feed and effluent concentrations of the spike were measured periodically and C/Co calculated. The C/Co values are plotted as a function of total feed solution volume in Figure 8. This curve shows no indication of plutonium breakthrough. The high C/Co values for this test compared to previous column tests with bone char are due to inefficiencies inherent in the small column. Fluctuations in the curve are caused by stops and starts in the feed flow.

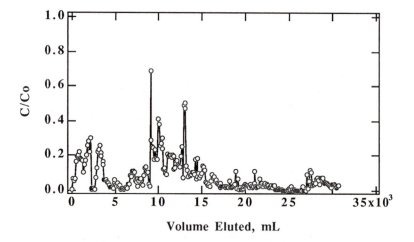

Figure 8. Plutonium Elution Curve for Small Column Adsorption Capacity Test

The required size of the process bone char columns for PFP wastewater can be estimated from adsorption rate measurements. Values of C/Co are plotted in Figure 9 for the 26 mL bone char columns as a function of residence time. The required C/Co value is defined as follows:

$$\frac{C}{Co} = \frac{\text{Maximum Activity in Effluent}}{\text{Estimated Activity in Feed}} = \frac{1.2 \text{ x } 10^{-6} \text{ μCi/L}}{4.3 \text{ x } 10^{-4} \text{μCi/L}} = 0.0028$$

where 1.2 x 10^{-6} μCi/L is the drinking water standard. In Figure 9 this C/Co value corresponds to a column residence time of about two minutes. Since the wastewater flow rate at the PFP is about 30 gal/min and the porosity of bone char is about 50%, the total required volume is 120 gallons.

Conclusions

Important characteristics of the ten adsorbents evaluated in this study relevant to plutonium removal from PFP wastewater are summarized and compared in Table IV. Properties that an ideal adsorbent should possess are (1) high adsorption distribution

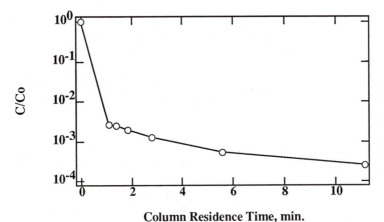

Column Residence Time, min.

Figure 9. Plutonium Adsorption on Bone Char for Various Column Residence Times

Table IV. Comparison of Adsorbents for Plutonium Removal from PFP Wastewater

Adsorbent	Equilibrium Pu Adsorption Kd, (mL/g)	Rate of Pu Adsorption	Comments
Alumina-C	30 to 50,000	NM	Adsorption strongly dependent on pH
MLM-300	8 to 16,000	NM	Adsorption strongly dependent on pH
Bone Char	8,300 to 31,000	Rapid	Adsorption maximum at pH 8.0
Amberlite IRC-718	60 to 200	Rapid	Small amount of Pu inert to resin
Duolite GT-73	160 to 3,200	Fairly rapid	Small amount of Pu inert to resin
Lewatit MP-500-FK	0.3 to 100	NM	Significant adsorption only at pH > 7
Dowex 21-K	0.1 to 17	NM	Significant adsorption only at pH > 7
Dowex M-41	0.5 to 510	NM	Significant adsorption only at pH > 7
Bioclaim-MRA	2,400 to 6,700	Slow	Adsorption not dependent on pH
Remsorb-CP	80 to 5,000	NM	Adsorption strongly dependent on pH

coefficients for plutonium over the entire range of pH expected for the wastewater, (2) rapid plutonium adsorption under wastewater conditions, (3) reactive toward all plutonium species that have significant concentrations in the wastewater, (4) large enough particle size to minimize pressure drop across columns containing the adsorbent, and (5) high mechanical strength and chemical stability. Of the ten adsorbents studied, only bone char appears to have all of these favorable properties. Each of the other adsorbents fail to meet at least one of the five criteria given above.

Literature Cited

1. Roberts, R. C. and W. R. Herald, *Development of Ultrafiltration and Adsorbents: April-September 1979*, MLM-2684, Monsanto Research Corporation, Mound Laboratory, Miamisburg, Ohio, 1980.
2. Silver, G. L. and J. W. Koenst, *A Study of the Reaction of Uranium and Plutonium with Bone Char*, MLM-2384, Monsanto Research Corporation, Mound Laboratory, Miamisburg, Ohio, 1977.
3. Blane, D. E. and E. L. Murphy, *Mound Laboratory Activities on the Removal of Plutonium and Uranium from Wastewater Using Bone Char*, MLM-2371, Monsanto Research Corporation, Mound Laboratory, Miamisburg, Ohio, 1976.
4. Smith, C. M., J. D. Navratil, and P. MacCarthy, "Removal of Actinides from Radioactive Wastewaters by Chemically Modified Peat," *Solvent Extraction and Ion Exchange*, **1984**, vol 2(7&8), pp. 1123-1149.
5. Koenst, J. W. and W. R. Herald, *Evaluation of New Macroporous Resins for the Removal of Uranium and Plutonium from Waste Streams*, MLM-2320, Monsanto Research Corporation, Mound Laboratory, Miamisburg, Ohio 1976.
6. Schulz, W. W. , J. W. Koenst, and D. R. Tallant, In *Actinide Separations*; Editors, J. D. Navratil and W. W. Schulz, ACS Symposium Series 117, American Chemical Society, Washington, D. C., 1980.
7. Nitsche, H., In *Material Research Society Symposium Procedings*; Material Research Society, 1991, Vol. 212, pp 517-529.
8. Allard, B. and J. Rydberg, In *Plutonium Chemistry*; Editors,W. T. Carnall and G. R Choppin, ACS Symposium Series 216, American Chemical Society, Washington D. C., 1983, pp. 275-295.
9. Grebenshchikova, V. I. and Yu. P. Davydov, *Soviet Radiochem.*, **1961**, vol 3, (2), 167.
10. Lindenbaum, A. and W. Westfall, *Int. J. Appl. Radiat. Isotop.*, **1965**, vol 16, 545.
11. Ockenden, D. W. and G. A. Welch, *J. Chem. Soc.*, **1956**, 3358-63.
12. Kinniburgh, D. G. and M. L. Jackson, In *Adsorption of Inorganics at Solid-Liquid Interfaces*; Editors, M. A. Anderson and A. J. Rubin, Ann Arbor Science Publishers Inc., Ann Arbor, Michigan, 1981, pp. 91-160.
13. Lindsay, W. L., 1979, *Chemical Equilibria in Soils*; John Wiley & Sons, New York, 1979, pp. 181.
14. Mailen, J. C., D. O. Campbell, J. T. Bell, and E. D. Collins, *Americium Product Solidification and Disposal*, ORNL/TM-10226, Martin Marietta Energy Systems, Inc., Oak Ridge National Laboratory, Oak Ridge, Tennessee, 1987.

RECEIVED January 24, 1992

Chapter 4

Decontamination of Groundwater by Using Membrane-Assisted Solvent Extraction

Joseph C. Hutter and G. F. Vandegrift

Separation Science and Technology Section, Chemical Technology Division, Argonne National Laboratory, 9700 South Cass Avenue, Argonne, IL 60439

Development of a new process to remove volatile organic compounds (VOCs) at dilute concentrations from groundwater was recently begun at Argonne. This process consists of membrane-assisted solvent extraction and membrane-assisted distillation stripping (MASX/MADS). The use of membranes in the solvent extraction and distillation stripping units improves mass transfer by increasing the interfacial surface area, thus allowing the process to be performed efficiently at high throughputs. The thermodynamics and regions of applicability of this process are discussed.

Contamination of groundwater by dilute concentrations of VOCs has become a major problem at many Department of Energy and Department of Defense sites, as well as industrial sites nationwide. Remediation of these sites is generally difficult, energy intensive, and very expensive. Environmental regulators only allow remediation procedures which can be proven to be less damaging to the environment than the original contamination; pump-and-treat technology, such as the MASX/MADS process described here, is usually acceptable to regulators. The MASX/MADS technology is also applicable to treatment of process air and water streams.

Groundwater contamination at landfills and chemical spill and dump sites can be so severe that two-phase liquid mixtures appear in the aquifer. In this situation, the organic contaminants are at their highest concentration, determined by each contaminant's equilibrium solubility. As the groundwater migrates through the aquifer, the contaminants are diluted. Often the contamination is in the ppb range only a few hundred yards from the dump site. Past practices, such as years of classifying now known carcinogens as "assorted non-toxics," have led to the decades of chemical dumping, which has caused the current environmental disaster. Over the decades, these contaminants have had time to be diluted by subsurface groundwater flows, and plumes have spread miles from the original contamination source.

Any remediation procedure must first prevent the source of contamination from adding to the problem. This means closing the dump site and eliminating the practices which led to the contamination in the first place. The near-site cleanup usually involves excavation, venting, or bioremediation to remove the high concentrations of organics in the unsaturated zone and two-phase liquid regions and thereby prevent further subsurface contamination spreading because of groundwater flows or intermittent rainfall. The unrecovered contaminants migrate through the aquifer with the groundwater flow and are retained by the adsorptive properties of the soil. Some degradation of the contaminants can occur in the subsurface environment. The contaminated groundwater is by far the most difficult problem to remediate due to its large extent in the subsurface environment. In general, the most widely used strategy to decontaminate the groundwater is to pump it from the aquifer using a well and then air strip the contaminants from the water. The decontaminated water can then be reintroduced to the environment (*1*).

A groundwater contamination survey was recently completed at nine DOE sites in the United States (Morrisey, C. M., et. al., Oak Ridge National Laboratory, unpublished data). The ten most common VOC contaminants found are listed in Table I. Most of the contaminants are chlorinated hydrocarbons, and they were typically found in concentrations in the ppb range. Similar contaminants can be found at the Argonne site (*2*) as well as several industrial sites worldwide (*3-8*). The common feature at all these sites is the large distances that dilute (ppb) concentrations of contaminants have migrated. Even though the concentrations of the contaminants are low, they often exceed US EPA Drinking Water Standards as shown in Table I (*9*). One of the contaminants shown in Table I, methylene chloride, is not under federal regulation, but it is usually under state or local restrictions.

Table I. Common Contaminants in Groundwater Found at DOE Sites

Contaminants	Average Concentration (ppb)	Maximum Concentration (ppb)	US EPA Drinking Water Standard (ppb)
Benzene	1400	36000	5
Carbon Tetrachloride	280	10000	5
cis-1-2-Dichloroethylene	100	1600	70
trans-1-2-Dichloroethylene	160	1600	100
Trichloroethylene	310	8500	5
Tetrachloroethylene	150	2300	5
Methylene Chloride	220	6400	NS[a]
1-1-1-Trichloroethane	110	3100	200
Chloroform	30	1000	100
Vinyl Chloride	180	540	2

[a]No US EPA Standard

Current Technology

Any remediation procedure must be able to reduce these low concentrations of contaminants in groundwater to the very dilute concentrations of the US EPA Standards. Solvent extraction in its conventional form is not adequate for this remediation procedure, since this technology is applicable to contaminations approaching the weight percent range and not the ppb range. The main difficulty with solvent extraction for wastewater treatment is regeneration of the solvent for recycle to the extraction section. This regeneration is usually done by conventional distillation. There are numerous literature references discussing the use of solvent extraction in wastewater treatment (*10-13*). The extraction step is usually relatively easy once a good solvent is found. Solvent extraction technology has led to the development of very effective solvent selection strategies based on either hydrogen bonding or Lewis acid-Lewis base concepts (*14-19*).

For groundwater applications the most common remediation procedure is air stripping followed by gas-phase activated carbon adsorption. Liquid-phase activated carbon adsorption is also used for removal of VOCs from water, but it is usually more expensive than the air stripping method. Ultraviolet (UV) radiation, catalyzed hydrogen peroxide or ozone is used to decontaminate groundwater by the destruction of the pollutants (*20*). Oftentimes, this technique cannot be used if the water is high in colloidal solids, which inhibit the oxidation reactions and/or foul the UV lamps. In-situ methods such as vacuum extraction or direct hydrogen peroxide injection to induce bioremediation have been applied in limited cases. In-situ bioremediation techniques are being researched, but so far have had only limited success. The major limitation of the in-situ bioremediation technologies is delivering oxygen and nutrients to the subsurface contaminants in so that the conditions required for biological activity can be maintained. In addition, the contaminants must be at concentrations high enough to support biomass. These limitations have confined bioremediation applications to very specific contaminants in highly contaminated regions, not the dilute regions of groundwater contamination usually treated by air stripping (*21*).

MASX/MADS Process

A flow diagram of the MASX/MADS process is given in Figure 1. Contaminated groundwater enters the membrane extraction module, where it is contacted with a solvent stream that extracts the pollutants from the groundwater flowing countercurrent to it . The membrane material in the module is porous and is used to separate the two liquid phases and provide a large interfacial area for mass transfer. The nonvolatile solvent extracts the VOCs from the water with a distribution coefficient that is typically 100 or more. Because of this high distribution coefficient, the oil flow rate can be up to 100 times lower than the groundwater flow. When the MASX module is designed properly, the groundwater leaves the extraction unit decontaminated to a purity as good as the drinking water standards, so that this water can be safely reintroduced to the environment. At the exit of the extraction module, the nonvolatile solvent contains the contaminants in a concentrated form recovered from the groundwater. The maximum concentration of the contaminants in the nonvolatile solvent is determined from the distribution coefficient between the water and the solvent, their relative flow rates, and the number of theoretical stages in the extraction module.

The solvent must be stripped of contaminants before it can be recycled to the extraction modules. This operation is carried out in the MADS unit after the solvent is heated. In this unit, the volatile contaminants are vaporized and recovered in a condenser. In addition to the contaminants, some of the solvent and residual groundwater dissolved in the solvent is recovered in the condenser. The decontaminated solvent is then recycled to the extraction modules. This process is not a conventional distillation step involving multiple stages; it is a one-stage evaporation, but stages may have to be added to (1) purify the nonvolatile solvent, or (2) treat the condensate if higher concentrations of VOCs are required for destruction or disposal.

In the MASX/MADS process, the use of membrane modules avoids any capacity or scaleup difficulties. Capacity is increased by adding more modules. Scaleup is easy because the well-defined interfacial area allows excellent characterization of the volumetric mass transfer rates. The modules also allow independent variation of the flow rates without flooding, as in conventional solvent extraction. Density differences between the solvent and extraction feed are not required for phase separation because the membrane separates the phases (*22*). Unlike

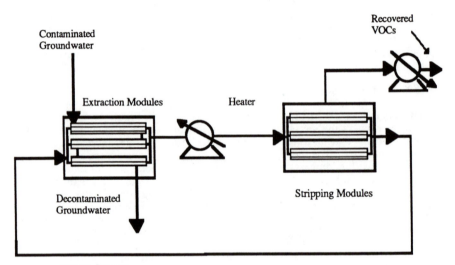

Figure 1. Flow diagram for the MASX/MADS process.

use of activated carbon with air stripping, the solvent is continuously regenerated and does not have to be removed to another process (steam stripping)to regenerate the adsorbent. Since this energy-intensive step is omitted, and the large volume of air required for air stripping does not have to be continuously blown through a packed column, this system should result in reduced energy consumption for operation.

Membrane Technology. Solvent extraction applications use microporous membranes with pore sizes that range from 0.005 μm to 0.1 μm. This pore size range overlaps ultrafiltration and microfiltration membranes. These membranes allow mass transfer to occur by diffusion which is only inhibited by tortuous pores. If a liquid is in the pores, mass transfer through the membrane is due to diffusion of material through the contained liquid phase. If a gas is in the pores, mass transfer is more rapid since diffusion is faster through a gas phase. The pore size and the surface energy of a membrane-contained fluid system will determine if the pores are gas filled or liquid filled for a given application. The transmembrane pressure is related to the surface tension and contact angle for a cylindrical pore by:

$$P = \frac{2 \, \sigma \cos \theta}{r} \tag{1}$$

Where P is the pressure (dynes/cm^2), σ the surface tension of the liquid (dynes/cm), θ is the contact angle (degrees), and r is the pore radius (cm) (27). This equation is for the idealized cylindrical pore case, and membrane structure can be more complicated than this simplified base case (34). The range of pore sizes and materials for available filtration membranes will allow a variety of configurations for new interface-immobilized solvent extraction applications. The design of the extraction module will be under the constraints determined by the pore sizes and materials. An immobilized liquid interface will occur only with the proper operating conditions. The significant differance for this membrane application is that the membranes are porous. Other membranes such as those used in gas separation or reverse osmosis, rely exclusively on mass transfer by a sorption-diffusion mechanism through a dense polymer above its glass transition temperature (23). Composite membranes have also been used in solvent extraction applications (24-26).

Due to the variety of membrane materials, configurations, and solvents available, mass transfer is controlled by several different mechanisms. For example, qualitative concentration profiles for extraction of a VOC from water by an organic phase are depicted in Figure 2. Experimental data for each of these mechanisms and detailed expressions for the overall mass transfer coefficient are reported in the literature (24-26). The thick solid lines in Figure 2 indicate a qualitative concentration profile. For each of these cases, a series of resistances determines the overall mass transfer rate. Using the film theory, a static boundary layer exists next to the membrane surface and the bulk fluid is in turbulent flow. The resistance to mass transfer includes boundary layer resistances on the aqueous and organic sides of the membrane as well as the resistance of the membrane itself. Typically, mass transfer is controlled by the resistances in one or more of these boundary layers, and the membrane resistance is a smaller contributing factor. For the extraction of VOCs from groundwater, the aqueous boundary layer has been identified as the major resistance to mass transfer using hollow fibers when a solvent with a high affinity for the VOCs has been used (24).

As depicted for the cases in Figure 2, the phase that preferentially wets the pores can be determined from free energy considerations of the system. A phase will wet a porous surface if the free energy of the wetted surface is lower than the sum of the free energies for the original liquid surface and the original porous surface. A

Hydrophobic Membrane

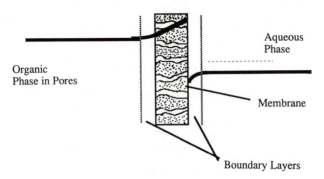

Hydrophilic Membrane

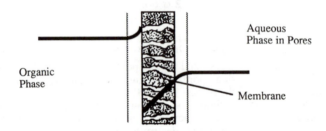

Composite Hydrophilic–Hydrophobic Membrane

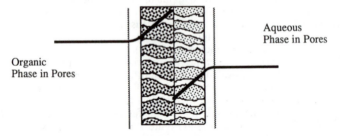

Composite Hydrophobic–Dense Membrane

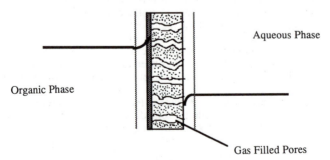

Figure 2. Concentration profiles in hydrophobic and hydrophilic and composite membranes used in a groundwater extraction module.

perfectly wetted surface has a contact angle of 0°. A partially wetted material has a contact angle between 0° and 90°. If the contact angle is above 90°, the material is not wetted. The exact contact angle of the system is determined from a balance of the liquid-solid, liquid-air, and solid-air surface energies. Only the liquid-air surface energy (or surface tension) is easily measured experimentally, so the complex thermodynamics of this system cannot be predicted a priori. Only qualitative arguments can be made with the limited available data, and the surface tension arguments given here are based on these limited data to predict trends for wetting behavior. Even though solid-air and liquid-solid surface energies are not known exactly, some useful trends can be predicted based on estimates of their magnitudes.

Figure 2a depicts a hydrophobic membrane. In this case the organic phase is in the pores, and the sum of the capillary force in the pores and the aqueous-phase hydrostatic pressure is higher than the organic-phase hydrostatic pressure; this prevents the organic phase from flowing out of the pores into the aqueous side of the membrane. Since mass transfer is occurring from the aqueous phase to the organic phase in this extraction, the concentration profile is decreasing from left to right in this figure. The discontinuity in the concentration profile is due to the distribution coefficient for this system. The discontinuity results from the equilibrium condition at the immobilized interface. If the pressure on the aqueous side of the membrane exceeds a critical value, the aqueous phase will be forced into the pores. This minimum pressure to wet the pores can be estimated by a force balance for a cylindrical pore (*27*).

In Figure 2b the aqueous phase is in the pores in a hydrophilic membrane. This is similar to the case in Figure 2a, except that the concentration gradient is greatest in the aqueous phase in the pores. For the case depicted in Figure 2b, the hydrostatic pressure of the aqueous phase is great enough to fill the pores preferentially to the organic phase, but it is lower than the sum of the organic phase hydrostatic pressure and capillary forces, which prevents the aqueous phase in the pores from flowing out of the pores into the organic phase.

The aqueous phase will wet the pores only if the free energy of the system is minimized compared to the free energy of the organic wetted surface. Hydrophilic materials such as ceramics or cellulose materials have high free surface energies, up to several thousand dynes/cm (ergs/cm^2). If the total surface energy of the aqueous wetted membrane is lower than that of the organic wetted membrane, the aqueous phase will preferentially wet the membrane. Water has a surface tension of 72 dynes/cm (or less with some impurities); thus, by wetting the hydrophilic surface the net surface energy is lowered (72 dynes/cm is less than several thousand dynes/cm). Polymer surfaces have a much lower surface energy, approaching 20 to 35 dynes/cm; thus, water will not wet many polymers since the net energy of the system will not be lowered. Wetting a low energy surface at 20 to 30 dynes/cm with a high surface energy liquid such as water at 72 dynes/cm will not lower the energy of the system. Instead, a small area of the surface will be in contact with water droplets having contact angles above 90°. Organics have surface tensions in the range of 20-30 dynes/cm; therefore, many more low energy surfaces can be wetted with organics than with water, since by wetting the porous surface with a low surface energy organic, the porous air-solid area, which has slightly higher energy, is reduced and the net energy of the system is lowered. It is also true that any surface wetted by water can be wetted by a lower surface tension organic phase as well (*28*).

A composite hydrophilic-hydrophobic membrane is shown in Figure 2c. The hydrophilic membrane is wetted by the aqueous phase, and the hydrophobic membrane is wetted by the organic phase. The interface is immobilized between the two membranes.

Figure 2d shows a composite hydrophobic-dense phase membrane. An example of such a membrane is a Celgard X-20 polypropylene hollow fiber porous

membrane with a 1-2 μm dense plasma polymerized disiloxane skin on the outside of
the fiber. This type of membrane has been used to demonstrate the solvent extraction
of contaminants from groundwater (24,29). In this lab-scale demonstration
membrane, the water in the lumen does not wet the hydrophobic surface, allowing the
pores to be gas filled. The pressure to wet this pore is given by equation 1. Thus, the
maximum length of a gas filled porous fiber is determined by the hydrostatic pressure
in the lumen. The pressure in the lumen is controlled by the fluid properties, fiber
diameter, and the fluid velocity in the lumen. Once the critical pressure given by
equation 1 is exceeded, the fiber pores will be liquid filled. On the outside of the
fibers, the dense skin prevents the organic phase from contacting the water or filling
the pores. The thin dense film permits rapid transfer of VOCs from the gas-filled
pores, allowing absorption by the organic phase. The organic phase is prevented from
contaminating the water, thus keeping the TOC (total organic carbon) low and meeting
strict environmental regulations.

Membrane Module Technology. The earliest membrane modules were the plate-
and-frame type. These modules are easy to clean if they are fouled because they can
be readily disassembled, but they do not have the high surface area per unit volume of
more advanced designs. The spiral-wound membrane initially used in reverse
osmosis applications consisted of a membrane sock with the open end connected to a
center pipe. Flow was through the sock surface, and permeate exited in the center
tube. The sock was spirally wound and placed in a pressure vessel shell. Millipore
Corporation has introduced a spiral-wound crossflow module for filtration
applications (30). Numerous commercial hollow fiber shell-and-tube modules give
the highest surface area to unit volume ratio. Tubular membranes also have the shell-
and-tube configuration. The main difference between tubular membranes and hollow
fibers is that the diameters of the tubes are considerably larger (1 cm). Most solvent
extraction laboratory research has been carried out with plate-and-frame modules or
hollow fiber modules. We are currently investigating the use of hollow fiber modules
for the MASX/MADS application.

Design of the MASX/MADS Process

Extraction Section. Since the concentrations of VOCs in groundwater must be
reduced, in most cases by several orders of magnitude, the extraction unit requires a
solvent with a very high affinity for the VOCs to minimize the solvent volume. The
affinity of a solvent for a VOC is defined by the distribution coefficient of each VOC:

$$D = \frac{C_{oil}}{C_{water}} \qquad (2)$$

Where D is dimensionless, and the liquid VOC concentrations, C_{oil} and C_{water}, can
have units of mol/L or mass/L. Large distribution coefficients enhance the extraction
performance. A lab-scale extraction module has been demonstrated for a synthetic
groundwater containing 1 ppm VOCs using a solvent with high VOC distribution
coefficients (24). This work has delineated a functional membrane-assisted solvent
extraction process to decontaminate water using solvents with distribution coefficients
for VOCs of 100 or more to minimize solvent volume and effectively operate the
extraction process. This process can be designed to achieve drinking water standards
(29). Because of the high distribution coefficient, the flow rate of the solvent can be
significantly lower than the flow rate of the groundwater. The large difference in flow
rates does not lead to poor stage efficiency in the membrane modules because of the
good interfacial contact. This large flow rate difference would make conventional
solvent extraction in a packed column or mixer-settler ineffective due to the poor stage

efficiency. In addition to membranes, the Argonne-design centrifugal contactor also has a high stage efficiency at organic-to-aqueous (O/A) flow ratios as low as 1/100 (*31*).

The design of the extraction unit depends on the distribution ratios, the decontamination factor, and the preferred flow rates of the groundwater and the solvent. For example, if chloroform (D=100) in groundwater were entering the unit at 1000 ppb, and the outlet chloroform concentration was designed to be less than or equal to 100 ppb (a decontamination factor of 10), the O/A flow ratio for one theoretical stage would need to be greater than or equal to 0.09. For two theoretical stages, only an O/A ratio of 0.026 would be required to give the same performance. Multiple theoretical stages in a countercurrent design would allow a considerably lower O/A ratio. These effects will be studied in future module designs.

Stripping Section. A high distribution coefficient for extraction of a VOC makes for easy decontamination of the groundwater and difficult stripping of the nonvolatile solvent. However, to be successful, the solvent must not only be a good extractant, but also easily regenerated. Removal of the VOC by air stripping is defined by Henry's Law defined below for an aqueous solution:

$$P \, y_G = H_{aq} \, C_{water} \tag{3}$$

Where H_{aq} is Henry's Law constant for a VOC in water, atm/unit concentration in the aqueous phase; P is the total pressure, atm; and y_G is the VOC mole fraction in the gas phase. The product $P \, y_G$ is the partial pressure of the VOC above the liquid phase. Henry's Law constants are specific to both the solute and the solvent. The higher the value of H_{aq}, the easier it is to remove the VOCs from the aqueous phase. In the nonvolatile solvent, the Henry's Law constant of the VOC ($H_{oil} = H_{aq}/D$) is lowered relative to water due to the greater affinity of the organic solvent for the VOC as measured by the large distribution coefficient. Substituting equation 2 into equation 3 yields:

$$P \, y_G = H_{aq} \, C_{water} = H_{oil} \, C_{oil} = \frac{H_{aq}}{D} \, C_{oil} \tag{4}$$

For a given concentration of VOC in the oil phase, the partial pressure of the VOC is lowered by a factor of D relative to the water case. For example, 1 ppm of chloroform in water has a partial pressure of 3.2×10^{-5} atm at 20°C. Because the value of D for chloroform is 100 between water and sunflower oil, it takes 100 ppm of chloroform in sunflower oil to obtain this same partial pressure.

For the intended application of MASX/MADS, ther are numerous candidates for solvents, as well as a range of possible contaminants in the groundwater. Table II gives the distribution ratios and Henry's Law constants for three selected contaminants treated with a water-sunflower oil system at 20°C. This table was generated from published data.(*24*)

Table II. Water-Sunflower Oil System at 20°C

Compound	D	H_{water} (atm/mol/l)	H_{oil} (atm/mol/l)
Chloroform	100	3.82	0.0382
Carbon Tetrachloride	867	30.2	0.0349
Tetrachloroethylene	2567	18.0	0.0070

As shown in Table II, chloroform and carbon tetrachloride have about the same Henry's Law constant above oil, even though carbon tetrachloride is considerably more volatile above water than chloroform. The high distribution coefficient for carbon tetrachloride significantly reduces the Henry's Law constant for the oil phase. The relatively high partial pressure of tetrachloroethylene above water is significantly reduced in oil due to the large distribution coefficient.

Regeneration of the Solvent, Ideal Inert Stripping-Gas Case

One way to regenerate the solvent is to strip the VOC contaminants using an inert stripping gas. The temperature and pressure in the stripping unit determine the gas flow rate, since Henry's Law determines the maximum VOC composition exiting in the strip gas. The partial pressures of 100 ppm for the three contaminants above sunflower oil at 20°C are given in Table III.

Table III. Partial Pressures of Three Candidate Contaminants[a]

Compound	Partial Pressure, Py_G (atm)
Chloroform	3.2×10^{-5}
Carbon Tetrachloride	2.9×10^{-5}
Tetrachloroethylene	0.43×10^{-5}

[a]100 ppm in Sunflower Oil, 20°C, 1 atm.

In situations where a combination of contaminants is involved, the least volatile component (in the case of Table III, tetrachloroethylene) controls the design of the stripper. Once the operating temperature and pressure is set, the exiting mole fraction of tetrachloroethylene is determined from its partial pressure. The mole fraction of the strip gas and its minimum flow rate can then be calculated.

    The design of the process has to account for several factors on a case by case basis. For example, for contamination by the three components listed in Table II, the extraction unit design is controlled by two factors: (1) the difficulty in extracting each component and, (2) the extent each components concentration must be decreased in the groundwater. Chloroform is the most difficult component to extract because it has the smallest distribution coefficient. However, if the attainment of EPA drinking water standards (factor 2) is the criterion, carbon tetrachloride with a D-value of about nine times that of chloroform with an EPA limit 1/20 that of chloroform is the most important. Depending on the extent of contamination and clean-up goals of the process, any of the other components could be the limiting factor in the design.

    The regeneration of the nonvolatile solvent is also controlled by two factors: (1) the difficulty in volatilizing each component above the nonvolatile solvent and, (2) the extent each component must be removed from the nonvolatile solvent. The volatization is related to the contaminant's partial pressure above the oil phase. This property depends on the contaminants concentration and its Henry's Law constant. The most difficult component to extract does not necessarily have to be the most difficult component to strip from the oil. The data in Table II indicate that at 20°C tetrachloroethylene will be the most difficult component to remove from the oil due to its low Henry's Law constant in the oil. Although a substantial amount of tetrachloroethylene can accumulate in the recycled nonvolatile solvent before the extraction performance exceeds its design value, the stripper must continuously remove enough tetrachloroethylene from the solvent with each pass so that the solvent can be recycled. These factors must be accounted for in the final design.

The economics of operating an air stripper to remove VOCs from water is controlled by the costs to run the blowers to deliver the air stream to the packed column stripper (*32*). The air requirement for removing VOCs from a nonvolatile solvent or water is set by Henry's Law and is the same whether a packed column or membrane module is used. Using membrane modules instead of packed columns for air stripping increases the compression power cost significantly since the modules are much smaller than the packed column and the gas phase pressure drop is increased. However, instead of using large gas volumes and compressor power, heat energy can be introduced to remove the VOCs from the nonvolatile solvent. At higher temperatures, the air volume requirement is significantly reduced because the volatility of the VOCs is increased. It is not practical to heat large volumes of water, but it may be practical to heat smaller volumes of nonvolatile solvents with much lower specific heat capacities compared to water.

Recovery of the VOC Components in the Condenser. To recover a VOC component in the condenser, the partial pressure of the VOC in the strip gas must exceed the vapor pressure of the condensed phase:

$$P \, y_i > P^{vap} \text{ to Condense} \qquad (5)$$

In an idealized stripping-gas case, only the VOCs are in the gas phase and should condense since the solvent is considered nonvolatile compared to the VOCs. For a 100 ppm chloroform concentration in sunflower oil at 20°C, the partial pressure of the chloroform (3.2×10^{-5} atm) does not exceed its vapor pressure at -60°C, so chloroform alone will not condense at this temperature, as indicated by the data in Table IV. By stripping the solvent at elevated temperatures, the VOC's partial pressure will increase and it should be possible to recover them in a condenser.

Table IV. **Vapor Pressures of Chloroform in the Condenser**[a]

Temperature (°C)	P^{vap} (atm)
20	2.6×10^{-1}
0	8.0×10^{-2}
-20	2.6×10^{-2}
-40	6.2×10^{-3}
-60	2.5×10^{-4}

[a]Partial pressure of chloroform in the stripping gas is 3.2×10^{-5} atm at 20°C.

Regeneration of the Solvent, Distillation Case

In the usual application of solvent extraction for the recovery of dilute VOCs from water, the solvent is regenerated by conventional distillation. This process is applied to aqueous systems which have much higher concentrations of contaminants than the ppb concentrations found in groundwater. The distillation procedure requires collection of vapor-liquid equilibrium (VLE) data for the VOC-solvent system before the distillation system can be designed. Typical vapor-liquid equilibrium curves for these VOC-solvent systems are given in Figure 3. This represents the VLE behavior expected for the natural oils and contaminants planned for use in the MASX/MADS process.

The extremely dilute VOC region in Figure 3 is of interest for the MASX/MADS process. As seen from the data, the vapor and liquid compositions converge as the VOC becomes more and more dilute. Thus, as the VOC concentration decreases, producing a high purity VOC product becomes increasingly difficult. Generating a high purity VOC product from a conventional distillation column would require operation at nearly total reflux and be very energy intensive. This problem has limited conventional solvent extraction to much higher concentrated streams, where the recovered solvent is contaminated enough to be regenerated by conventional distillation, and a highly concentrated impurity product is produced as well. In the ppm or ppb region encountered for the MADS, the solvent vapor partial pressure is much higher than the VOC partial pressure. Because of this, by vaporizing and condensing some of the solvent along with the VOC and residual groundwater extracted by the solvent, it should be possible to recover an enriched solution of the VOC in the solvent. For example, a VOC solution of solvent may produce a condensate which is several orders of magnitude more concentrated then the original groundwater. The remaining solvent will have a substantially lower VOC concentration and can then be recycled to the extraction unit. Condensing the VOC-solvent vapor will be much easier since its dew point will be much higher than the required temperature to recover the VOCs directly from an inert strip gas, as discussed previously. Enriching the VOC vapor that is condensed to a greater extent will require additional stages, with each MADS stage smaller than the previous one due to the lower liquid flow rate obtained with each enriching step.

Advantages of Membranes. The mass transfer rate for this type of separation process is given by

$$J = K_L a \left(C - C^* \right) \tag{6}$$

Where $K_L a$ is the mass transfer coefficient, C^* is the VOC concentration in the solvent in equilibrium with the vapor phase, C is the actual solvent VOC concentration, and J is the flux of the VOC. Raising the temperature will cause C^* to approach zero. For the dilute VOC in the solvent, small values of C cause the flux to approach zero and separation does not occur. By using membranes with their large area to volume ratio, the mass transfer coefficient can be increased by an order of magnitude or more compared to a conventional packed column (*33*). This increase in area will enhance flux despite small values of C, thus making the separation more feasible. The VLE data needed to evaluate the MASX/MADS process are currently being collected. It is expected that this process will perform well.

Conclusions

Membrane-assisted solvent extraction of dilute VOCs from groundwater is technically feasible. The regeneration of the solvent is the difficult step in the development of the process. It is not feasible to air strip the solvent at ambient conditions due to the low partial pressures of the VOCs above the solvent. These low partial pressures in the strip gas make the VOCs impossible to recover with a condenser. However, by using

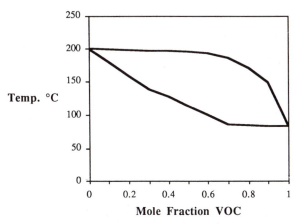

Figure 3. Vapor-liquid equilibrium for the VOC-solvent system.

a membrane-assisted vaporization procedure, the VOCs can be recovered from the organic phase. Some vaporization of the solvent will occur, and this two-phase system will condense at a dew point much higher than the required temperature to recover the VOCs directly from an inert strip gas.

Work supported by the U.S. Department of Energy, Separation Science and Technology Program, under Contract W-31-109-Eng-38.

Literature Cited

[1] Brown, R. A., Sullivan, K. *Poll. Eng.* **1991**, *23*, 62-68.

[2] Moos, L. P., Duffy, T. L. *1990 Annual Site Environmental Report*:; ANL-90/8, Argonne National Laboratory, Argonne, IL, 1990.

[3] Goodrich, J. A., Lykins, B. W., Clark, R. M., Oppelt, E. T. *J. AWWA* **March 1991,** 55-62.

[4] DeWalle, F. P., Chian, E. S. K. *J. AWWA* **April 1981,** 206-211.

[5] Stuermer, D. H., Ng, D. J., Morris, C. J., *Env. Sci. Tech.* **1982,** *16*, 582-587.

[6] Reinhard, M., Goodman, N. L., Barker, J. F. *Env. Sci. Tech.* **1984,** *18*, 953-961.

[7] Roberts, A. J., Thomas, T. C. *Env. Tox. Chem.* **1986,** *5*, 3-11.

[8] Forst, C., Roth, W., Kuhnmunch, S. *Int. J. Env. Anal. Chem.* **1989,** *37*, 287-293.

[9] U. S. Environmental Protection Agency, *Drinking Water Regulations and Health Advisories*, April 1990.

[10] Sandall, O. C., Shiao, S. Y., Myers, J. E. *AIChE Sym. Ser.* **1975,** *70*, 24-30.

[11] Hwang, S. T. *AIChE Sym. Ser.* **1981,** *77*, 304-315.

[12] Ricker, N. L., King, C. J. *AIChE Sym. Ser.* **1978,** *74*, 204-209.

[13] Hewes, C. G., Smith, W. H., Davis, R. R. *AIChE Sym. Ser.* **1975,** *70*, 54-60.

[14] Earhart, J. P., Won, K. W., Wong, H. Y., Prausnitz, J. M., King, C. J. *CEP* **May 1977,** 67-73.

[15] Ewell, R. H., Harrison, J. M., Berg, L. *Ind. Eng. Chem.* **1944,** *36*, 871-875.

[16] Joshi, D. K., Senetar, J. J., King, C. J., *Ind. Eng. Chem. Proc. Des. Dev.* **1984,** *23*, 748-754.

[17]Barbari, T. A., King, C. J. *Env. Sci. Tech.* **1982**, *16*, 624-627.

[18]King, C. J., Barbari, T. A., Joshi, D. K., Bell, N. E., Senetar, J. J. *Equilibrium Distribution Coefficients for Extraction of Chlorinated Organic Priority Pollutants from Water -1* EPA-600/2-84-060a; U. S. Dept. of Commerce, National Technical Information Service: Springfield, VA, 1984.

[19]King, C. J., Barbari, T. A., Joshi, D. K., Bell, N. E., Senetar, J. J. *Equilibrium Distribution Coefficients for Extraction of Chlorinated Organic Priority Pollutants from Water -2* EPA-600/2-84-060b; U. S. Dept. of Commerce, National Technical Information Service: Springfield, VA, 1984.

[20]Hager, D. G., *Physical/Chemical Processes Innovative Hazardous Waste Treatment Technology Series* ; Freeman, H. M., Ed., Innovative Hazardous Waste Treatment Technology Series-Volume 2; Technomic: Lancaster, PA, 1990, 143-154.

[21]Roulier, M., Ryan, J., Houthoofd, J., Pahren, H., Custer, F., *Physical/Chemical Processes Innovative Hazardous Waste Treatment Technology Series* ; Freeman, H. M., Ed., Innovative Hazardous Waste Treatment Technology Series-Volume 2; Technomic: Lancaster, PA, 1990, 199-204.

[22]Prasad, R., Sirkar, K. K. *AIChE J.* **1987**, *33*, 1057-1066.

[23]*Chemical Engineers Handbook,* Perry, R. H., Green, D. W., Eds., McGraw-Hill: Highstown, NJ, 1989, 17-14 - 17-34.

[24]Zander, A. K., Qin, R., Semmens, M. J. *J. Env. Eng.* **1989**, *115*, 768-784.

[25]Prasad, R., Sirkar, K. K. *Sep. Sci. Eng.* **1987**, *22*, 619-640.

[26]Prasad, R., Frank, G. T., Sirkar, K. K. *AIChE Symp. Ser.* **1988**, *84*, 42-53.

[27]Callahan, R. W. *AIChE Symp. Ser.* **1988**, *84*, 54-63.

[28]Rosen, M. J., *Surfactants and Interfacial Phenomena*, John Wiley & Sons, Inc.: New York, NY, 1978, 174-184.

[29]Semmens, M. J., *Method of Removing Organic Volatile and Semivolatile Contaminants from Water*, U. S. Patent # 4960520, 1990.

[30]Millipore, *Analysis, Purification, Monitoring, Quality Control,* 1990 Catalog.

[31]Leonard, R. A., *Chemical Technology Division Annual Report, 1989*, ANL-90/11, Argonne National Laboratory, Argonne, IL, 1990.

[32]Byers, W. D., *Physical/Chemical Processes Innovative Hazardous Waste Treatment Technology Series* ; Freeman, H. M., Ed., Innovative Hazardous Waste Treatment Technology Series-Volume 2; Technomic: Lancaster, PA, 1990, 19-29.

[33]Zander, A. K., Semmens, M. J., Narbaitz, R. M. *J. AWWA* **1989**, *81*, 76-81.

[34]Chaiko, D., Osseo-Asare, K. *Sep. Sci. Tech.* **1982**, *17*, 1659-1667.

RECEIVED January 24, 1992

Chapter 5

Low-Temperature Transportable Technology for On-Site Remediation

Michael G. Cosmos[1] and Roger K. Nielson[2]

[1]Roy F. Weston, Inc., West Chester, PA 19380
[2]SoilTech, ATP Systems, Inc., 6300 Syracuse Way, Suite 300, Englewood, CO 80111

An innovative technology, the Low Temperature Thermal Treatment (LT^3) System was developed by Roy F. Weston, Inc. to treat soils, sludges and sediments contaminated with volatile and semivolatile organic compounds. The LT^3 process was utilized to remediate a contaminated site near Crows Landing, CA. The processed soil was treated to less than 5 ppm of total petroleum hydrocarbons. During operation stack emissions were analyzed to determine emission characteristics.

Large quantities of soil sludges and sediments have been contaminated with low levels of organic chemicals as the result of leaking storage tanks, waste disposal activities and industrial or commercial operations. The treatment of large quantities of impacted soils is expensive when using existing thermal technologies such as incineration. To provide an effective and cost efficient method of treating organic contaminated soils Roy F. Weston, Inc. has developed and patented its Low Temperature Thermal Treatment (LT^3) System (U.S. Patent No. 4,738,206).

LT^3 Process Description

The basis of the LT^3 technology is the thermal processor, which is an indirect heat exchanger used to dry and heat contaminated soils. Heating the soils to approximately 500°F in the processor evaporates or strips the organics from the soil. The organic vapors are then vented through two condensers which operate

0097–6156/92/0509–0061$06.00/0
© 1992 American Chemical Society

in series to condense and remove organic compounds. The vapor stream is subsequently treated by carbon adsorption. Treated gases are discharged to the atmosphere. The condensate stream, which is primarily the inherent soil moisture, is also treated by carbon adsorption. Treated water is reused as cooling water in the soil conditioner or as dust control in excavation and backfill areas.

The LT^3 system is divided into three main areas of emphasis: soil treatment, emissions control, and water treatment. A schematic diagram of the LT^3 process is shown in Figure 1. The LT^3 process equipment is mounted on three tractor trailer beds for transportation and operation. The unit is suitable for highway transport. The general arrangement of the process equipment and the placement of the trailers during operation is shown on Figure 2.

Soil Treatment

Soil Feed System. Excavated soil is transported to the system by a front-end loader. The front-end loader carries the soil over a weigh scale. The soil weight is recorded for each load transported to the power shredder. Soil is deposited directly into the power shredding device. Classified soil with a top size of less than 2 inches passes through the shredder into the feed conveyor. Oversized material is removed and stockpiled in a roll-off container. The feed conveyor discharges into the surge hopper located above the thermal processor. The soil will be fed into the LT^3 system at regular intervals to maintain the surge hopper seal. The surge hopper also provides a seal over the thermal processor to minimize air infiltration.

Thermal Processor. The thermal processor consists of two jacketed screw conveyors. The shafts and flights of the screw conveyors and the trough jackets are hollow to allow circulation of a heat transfer fluid (i.e., hot oil). The function of each screw conveyor is to move soil forward through the processor and to thoroughly mix the material, providing indirect contact between the heat transfer fluid and the soil. The troughs are assembled in a piggyback fashion (one above the other). Each trough houses four intermeshed screw conveyors.

The screw conveyors are intermeshed to break up soil clumps and to improve heat transfer. The four screws of each processor are driven by a variable speed drive

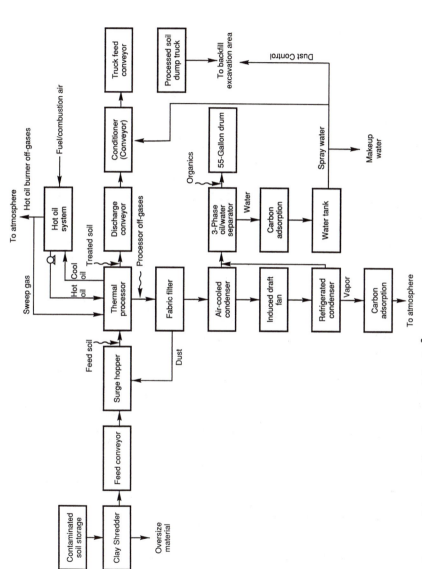

Figure 1. Schematic flow diagram of the LT³ process. (Reproduced with permission from Roy F. Weston, Inc.)

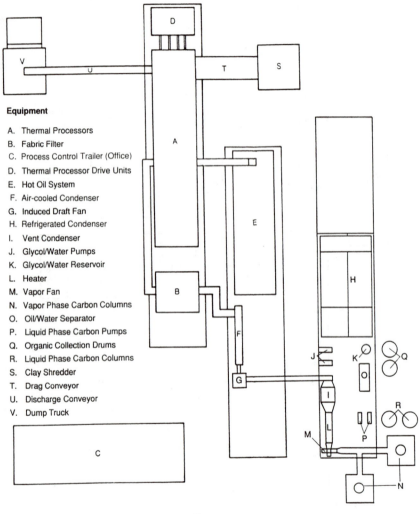

Equipment

A. Thermal Processors
B. Fabric Filter
C. Process Control Trailer (Office)
D. Thermal Processor Drive Units
E. Hot Oil System
F. Air-cooled Condenser
G. Induced Draft Fan
H. Refrigerated Condenser
I. Vent Condenser
J. Glycol/Water Pumps
K. Glycol/Water Reservoir
L. Heater
M. Vapor Fan
N. Vapor Phase Carbon Columns
O. Oil/Water Separator
P. Liquid Phase Carbon Pumps
Q. Organic Collection Drums
R. Liquid Phase Carbon Columns
S. Clay Shredder
T. Drag Conveyor
U. Discharge Conveyor
V. Dump Truck

Figure 2. General arrangement of LT^3 system. (Reproduced with permission from Roy F. Weston, Inc.)

mechanism. Residence time and soil temperature in the thermal processor are adjusted by varying the rotational speed of the screws and the hot oil temperature setting.

Vapors are driven off the soil and are drawn out of the thermal processor by an induced draft (ID) fan. The draft created by the ID fan is maintained in the processor to allow the vapors to be removed from the processor.

Processed Soil Handling System. Soil is discharged from the thermal processor into a horizontal screw conveyor. The horizontal screw conveyor discharges to a second screw conveyor, or ash conditioner. The ash conditioner is a ribbon flight screw conveyor. Water spray nozzles are installed in the ash conditioner housing to cool the discharged soil and to minimize fugitive dust emissions. The conditioner discharges onto an inclined stacker belt.

The stacker belt conveys the wetted processed soil from the conditioner to a dump truck.

Hot Oil System. The hot oil system is a self-contained unit consisting of a 7.2 million British thermal units per hour (Btu/hr) gas-fired burner, flame supervisory system, oil reservoir, hot oil pump and associated controls.

The hot oil system burner provides the thermal energy to maintain the temperature of the heat transfer oil used to indirectly volatilize organics from the soil.

A portion of the combustion gases released from the hot oil system is used as sweep gas in the thermal processor. The warm sweep gas (i.e., 700°F and very low oxygen content) removes the organics from the processor. Sweep gas is introduced to maintain an exhaust gas temperature from the processor of about 350°F. The sweep gas also provides an inert atmosphere to avoid exceeding the lower explosive limit (LEL) of contaminants within the thermal processor and downstream equipment. The remaining portion of combustion gases from the oil system is released for atmospheric discharge.

Emission Control System. An emission control system is provided to prevent the release of particulate matter or organic vapors. The control devices are described in this subsection and include the fabric filter, air-cooled condenser, induced draft (ID) fan, refrigerated condenser, and activated carbon system.

Fabric Filter. Sweep air and organics from the thermal processor are drawn by the ID fan into a fabric filter for particulate (dust) removal. The fabric filter is of the jet-pulse design, where air at a pressure of 80 pounds per square inch gauge (psig) periodically pulses to remove dust that has accumulated on the bags. Dust drops to the bottom of the fabric filter and is collected in two collection bins. Dust is manually removed from the fabric filter.

Air-Cooled Condenser. The exhaust gas from the fabric filter is drawn into an air-cooled condenser by the ID fan. The air-cooled condenser is used to remove condensible water vapor and organics from the exhaust gas. Condensed liquid is collected in a trap and is pumped to the water treatment system. Condenser off-gases are directed to the second (refrigerated) condenser.

Refrigerated Condenser. The function of the refrigerated condenser is to further lower the temperature and moisture content of the exhaust gases, thereby reducing the water/organic load on the vapor phase carbon system.

Saturated exhaust gas from the air-cooled condenser enters the refrigerated condenser at 125°F. Circulating glycol/water solution indirectly cools the process gas to 60°F. The glycol/water solution is continuously recycled through a refrigerator that maintains the glycol/water temperature.

The process gas is then reheated to 70°F by three electric resistance heaters. The relative humidity of the exiting gas is 70%, preventing further condensation in the process piping, and optimizing the activated carbon adsorption efficiency.

Induced Draft (ID) Fan. The ID fan is used to induce air flow through the thermal treatment system and prevent fugitive emissions of dust on volatile organics.

Vapor-Phase Carbon Adsorption System. The function of the vapor-phase activated carbon columns is to remove the remaining organics from the exhaust fumes exiting the refrigerated condenser.

Approximately 99% of the organics entering the carbon unit will be removed from the vapor stream.

The system includes two carbon columns. The vapor-phase carbon columns are operated in parallel to allow isolation and access to one column while the other is on-line. The gases exiting the carbon columns may contain total organic concentrations up to 4 parts per million (ppm) by volume.

The exhaust gas from the carbon columns is continuously monitored for total hydrocarbon (THC) concentrations. If the THC concentration exceeds 10 ppm, the carbon will be considered "spent" and an alarm will be activated.

Continuous Emissions Monitoring (CEM) System. An extractive-type continuous emissions monitoring (CEM) system is used to monitor the carbon column exhaust gases for oxygen, carbon dioxide, carbon monoxide and total hydrocarbons. The CEM system also monitors the exhaust gases from the fabric filter (air-cooled condenser inlet) for oxygen and total hydrocarbons.

Condensate Handling System. The condensate from the LT^3 system is separated into light and heavy organic compounds and water. The aqueous phase (water) is treated by carbon adsorption and recycled to the LT^3 process. The organic phases are disposed of at an off-site permitted facility. The condensate handling system is described in this subsection.

Oil-Water Separator. Liquid exiting both the air-cooled and refrigerated condensers is collected and pumped to a 3-phase oil-water separator allowing the insoluble light and heavy organic components to separate from the water. The light organic phase is removed by a skimmer and weir. The heavy organic phase is removed through a drain. The water phase flows out of the separator and is directed to the liquid-phase carbon adsorption system.

Carbon Adsorption. The water removed from the oil-water separator is directed through two carbon adsorption columns that operate in series for removal of soluble organics. The liquid stream between the two carbon columns is routinely sampled to detect breakthrough in the first carbon column.

Utilities. Operating the LT^3 system requires the following support systems and utilities:

- Electrical power
- Diesel fuel
- Natural gas
- Process water

Case Study

The LT^3 system was applied at a site near Crows Landing, CA to remediate 1,400 tons of soil. The soil was contaminated with fuel oil, jet fuel, and spent solvents. The soil was contaminated as the result of fuel and solvents spilled in the fire training practice area. The soil was impacted with BTEX, (Benzene, toluene, ethyl benzene and xylene) as well as total petroleum hydrocarbons. Site characterization data collected in 1988 indicated concentrations of TPH up to a maximum of 5,400 ppm. The characterization data was compiled from surface samples and soil borings in the impacted area. The average composition of the ten samples in the impacted area was 619 ppm. During processing, grab samples were collected from the untreated material as it was fed to the processor. The concentrations were nonuniform and range from 97 ppm to non detected (1 ppm detection limit). The average concentration of the grab samples was 52 ppm $\pm$ 32 ppm.

The LT^3 system was mobilized to the site after preparation of a detailed site specific Work Plan and Health and Safety Plan. An Air Permit was received from the Stanislaus County Air Resources Board. The soil was excavated from a 50 ft. by 50 ft. area. During treatment the treated soil was composited daily and analyzed using a Hanby Environmental Test Kit for petroleum hydrocarbons. This simple test kit, which provides rapid soil analysis, was used as a means of process control. The processed soil operating temperature and retention time was optimized at 422°F and 22 minutes, respectively. The treated soil samples were collected and analyzed for TPH and BTEX's by an independent third party. The average of the 18 samples collected and analyzed using approved analytical techniques are provided on Table I. The treated soil exceeded the treatment criteria of 100 ppm total petroleum hydrocarbons and 700 ppb toluene.

Table I. Average Treated Soil Results

Total Petroleum Hydrocarbons[a]	5.2 ± 8	ppm
Benzene[b]	< 5	ppb
Toluene[b]	18 ± 36	ppb
Ethyl Benzene[b]	< 5	ppb
Xylene[b]	< 15	ppb

a - TPH measured using CA DHS Luft Method

b - BTEX measured using EPA Method 8020

During the processing the Stanislaus County Air Resources Board established a 10 ppm total hydrocarbon emission criteria. The continuous emission monitor sampling the process gases at the discharge of the carbon adsorption column reached a maximum 9 ppm at which time a second carbon unit was engaged. A total of 4,400 lbs. of carbon were utilized during the treatment.

To further characterize the process emissions, three stack tests were conducted using EPA Volatile Organic Sampling Train (VOST) and EPA modified Method 5 for semivolatile organics including polynuclear aromatics. The stack test results of the volatile and semivolatile species indicated very minor emissions of detectable compounds. The total detected semivolatile compounds averaged 14.3 micrograms per cubic meter and consisted of three detectable compounds; phenanthrene, naphthalene and fluoranthene. The volatiles were detected at an average emission of 86 micrograms per cubic meter, 75% of the volatile emissions consisted of dichlorodifluoromethane and chloromethane commonly detected volatile organic compounds which were not identified as possible site contaminants.

The data from the Crows Landing Site remediation illustrated that the LT^3 system is an applicable technology for removal of petroleum contaminates from soil while meeting stringent air emission criteria.

RECEIVED April 24, 1992

Chapter 6

Use of Chelating Agents for Remediation of Heavy Metal Contaminated Soil

Robert W. Peters and Linda Shem

Energy Systems Division, Argonne National Laboratory, 9700 South Cass Avenue, Argonne, IL 60439

The efficiency of removing lead (Pb) from a contaminated soil of high clay and silt content was studied in a laboratory investigation where chelating agents were used to extract the lead from the soil. Lead concentrations in the soil ranged from 500 to 10,000 mg/kg. Ethylenediaminetetraacetic acid (EDTA) and nitrilotriacetic acid (NTA) were examined for their potential extractive capabilities. Concentrations of the chelating agents examined in this study ranged from 0.01 to 0.10 M. The pH of the suspensions in which the extractions were performed ranged from 4 to 12. Results showed that the removal of lead using NTA and water was pH-dependent, whereas the removal of lead using EDTA was pH-insensitive. For an initial lead concentration of 10,000 mg/kg, the maximum removals of lead using EDTA, NTA, and water (compared with nominal initial Pb concentrations) were 64.2, 19.1, and 5.8%, respectively.

Subsurface soils and groundwaters can become contaminated with heavy metal as a result of a number of activities, including the application of industrial waste, fertilizers, and pesticides; mining, smelting, and metal plating/metal finishing operations; automobile battery production; vehicle emissions; and fly-ash from combustion/incineration processes.

Metal concentrations in groundwater are largely governed by interactions with surrounding soils and geological materials. Many different mechanisms influence the partitioning of metals between the solid and solution phases (thereby affecting the leachability of metals from contaminated soils), including dissolution and precipitation, sorption and exchange, complexation, and biological fixation (1). The major effect of complexation is a dramatic increase in the solubility of the heavy metal ions, especially for strong complexing agents such as ethylenediaminetetraacetic acid (EDTA) and

0097–6156/92/0509–0070$06.00/0

nitrilotriacetic acid (NTA) (*2,3,4,5*). Because of this increased solubility of metal ions in solution, complexing agents offer the potential to effectively extract heavy metals from contaminated soil.

Background

Reported Heavy Metal Concentrations. Numerous investigators have reported concentrations of heavy metal in soils contaminated by various activities. Oyler (*6*) reported that concentrations of zinc in soils typically range from 10 to 300 ppm, while concentrations of zinc resulting from zinc smelting operations on Blue Mountain in Palmerton, Pennsylvania, ranged from 26,000 to 80,000 ppm. Similarly, concentrations of cadmium typically range from 0.1 to 7.0 ppm, while concentrations of cadmium reported on Blue Mountain ranged from 900 to 1,500 ppm. Lead has been reported in concentrations as high as 6,475 mg/L. Field samples collected from contaminated sites in northern New Jersey had chromium concentrations ranging from 500 to 70,000 ppm (*7*).

Several studies of soils at battery recycling facilities have also reported heavy metal concentrations. Trnovsky et al. (*8*) found that lead was the primary contaminant in soil, surface water, sediments, and groundwater at a former battery reclamation site in northern Florida, although elevated concentrations of antimony, arsenic, cadmium, and selenium were also present. The results of Trnovsky's analyses showed lead concentrations ranged from 2 to 135,000 ppm in soil samples, 2.16 to 34,700 ppm in the on-site river sediment samples (*8,9*), and 0 to 140 mg/L in surface waters (*10*). The lead concentration in the soil was influenced by various soil characteristics; the extent of contamination decreased with increasing soil depth. Wagner (*11*) noted that both lead and zinc migrated only up to 10 cm in a clay marl (with approximately one-third of the rock containing carbonate) under the driving force of diffusion and retardation, which were attributed primarily to precipitation. The contamination zone of lead is usually confined to the top one foot of soil because of strong metal-soil affinity.

In another study, Elliott et al. (*12*) noted that surface soils in the vicinity of lead smelting operations can have concentrations as high as 6.2%. Soils at former battery recycling sites contained an average lead concentration of 5% (*13*) and had reported lead concentrations as high as 211,300 mg/kg soil (dry weight) (*12,14*). Hessling et al. (*15*) investigated soil washing as a remediation technique for lead-contaminated soils at battery recycling facilities. Six sites were studied, with lead concentrations ranging from 210 to 75,850 mg/kg in soil. At some sites, elevated levels of other contaminants (such as cadmium, copper, nickel, arsenic, and zinc) were also detected.

The reported ranges of heavy metal concentrations listed in the technical literature are summarized in Table I.

High inputs of heavy metal to soils may result in the saturation of available surface sorption and complexation sites. Under these conditions, the soil serves as a "pollution source" rather than a "sink" (*1*). In addition to adding metal cations, which overwhelm the soil's retention capacity, anthropogenic activities may supply metals to soils in more geochemically mobile forms. High concentrations of metals in groundwater can result from the soil deposition of wastewaters, municipal and industrial wastes, and mining and ore processing residues (*1*). The movement of copper, cadmium, and nickel into a shallow aquifer from sewage oxidation ponds was attributed to organic chelation of the metals that were only weakly sorbed by coarse-textured soils (*16*). Microbial oxidation of sulfides can also cause heavy metals to be leached as groundwater passes through tail minings (*17*). Elevated groundwater concentrations of nickel and cadmium resulted from waste disposal activities at a battery factory in Sweden (*18*).

Remediation Techniques. The excavation and transport of soil contaminated with heavy metals has been the standard remedial technique. This technique is not a

Table I. Reported Heavy Metal Concentrations

Heavy Metal	Site Type	Reported Concentration Range (ppm)	Reference
Zn	Smelting	26,000-80,000	(6)
Cd	Smelting	900-1,500	(6)
Cr	Chromium production	500-70,000	(7)
Pb	Battery reclamation	2-135,000 (soil)	(8,9)
		2.16-43,700 (on-site sediment)	(8,9)
		0-140 mg/L (surface waters)	(10)
Pb	Battery recycling	210-75,850 mg/kg soil	(15)
		0-211,300 mg/kg soil	(12,14)

permanent solution; moreover, off-site shipment and disposal of the contaminated soil involves high expense, liability, and appropriate regulatory approval. Further, recent U.S. Environmental Protection Agency (EPA) regulations require pretreatment prior to landfilling (19). This requirement has stimulated interest in technologies that can be used to treat contaminated soils either on-site or in situ (1).

For soils contaminated with organic pollutants, a number of techniques can be considered for the remediation of a particular site, including thermal treatment, steam and air stripping, microbial degradation, and chemical oxidation. In contrast, few treatment techniques are available for the remediation of metal-laden soils. Metals can be removed by either flotation or extraction. Process parameters affecting extraction technologies for cleaning up soils have been summarized by Raghavan et al. (20). The migration of metals can also be minimized by solidifying or vitrifying the soil and fixing the metals in a nonleachable form. In solidification processes, lime, fly ash, cement-kiln dust, or other additives are added to bind the soil into a cement-like mass and immobilize the metallic compounds. In vitrification processes, the soil is formed into a glassy matrix by applying current across embedded electrodes. Solidification is very expensive, because the waste must be thoroughly characterized to determine its compatibility with the specific treatment processes. Many existing technologies result in a solid with unacceptable long-term stability. Solidification/stabilization also results in an increase in the volume of the waste. Following remediation, site reuse is limited, and long-term monitoring is generally required. For these reasons, solidification/stabilization is generally limited to radioactive or highly toxic wastes (21).

Techniques for removing metals from soil generally involve bringing the soil into contact with an aqueous solution. Methods such as flotation and water classification are solid/liquid separation processes. Metal contamination is generally found on the finer soil particles. Because metals are often preferentially bound to clays and humic materials, separating the finely divided material may substantially reduce the heavy metal content of the bulk soil. Separating the fines from the coarse fraction was found to remove 90% of the metals but only 30% of the total dredged material (22). In flotation processes, metallic minerals become attached to air bubbles that rise to the surface, thereby forming a froth, and they are therefore separated from particles that are wetted by water (23). Although flotation has been used successfully in the mineral processing industry, it has not been widely used for the treatment of contaminated soils (*1*). Solid/liquid separation processes represent a preliminary step in the remediation of contaminated soils. The metals are still bound to a solid phase, but in a much more concentrated form. The advantage of this preliminary processing is that the mass of contaminated soil to be processed is considerably reduced; therefore, the treatment cost is reduced.

Pickering (*24*) identified four approaches whereby metals could be mobilized in soils: (1) changes in acidity, (2) changes in system ionic strength, (3) changes in oxidation/reduction (REDOX) potential, and (4) formation of complexes. In the latter technique, the addition of complexing ligands can convert solid-bound heavy metal ions into soluble metal complexes. The effectiveness of complexing ligands in promoting the release of metals depends on the strength of bonding to the soil surface, the stability and adsorbability of the complexes formed, and the pH of the suspension (*1*). From an application viewpoint, the type and concentration of the complexing ligand and the system pH are the operational parameters that can be controlled.

The ability of chelating agents to form stable metal complexes makes materials such as ethylenediaminetetraacetic acid (EDTA) and nitrilotriacetic acid (NTA) promising extractive agents for the treatment of soils polluted with heavy metals (*25*). Elliott and Peters (*1*) have noted that although complexation is the major mechanism responsible for the metal solubilization, the overall release process depends on the hydrogen ion concentration and the system ionic strength. Because hydrous oxides of iron and manganese can coprecipitate and adsorb heavy metals, they are believed to play an important role in the fixation of heavy metals in polluted soils (*26*). Their dissolution under reducing conditions may weaken the solid heavy-metal bond and thereby promote solubilization of the metal ions. Elliott and Peters (*1*) noted that there are five major considerations in the selection of complexing agents for soil remediation:

1. Reagents should be able to form highly stable complexes over a wide pH range at a 1:1 ligand-to-metal molar ratio.

2. Biodegradability of the complexing agents and metal complexes should be low (especially if the complexing agent is to be recycled for reuse in the process).

3. The metal complexes that are formed should be nonadsorbable on soil surfaces.

4. The chelating agent should have a low toxicity and low potential for environmental harm.

5. The reagents should be cost-effective.

Elliott and Peters (1) note that although no compounds ideally satisfy all these criteria, there are several aminocarboxylic acids that form remarkably stable complexes with numerous metal ions. The properties of three chelating agents that have been used to extract contaminants from soil (either for analyses or remediation) are summarized in Table II.

Extraction of heavy metals from contaminated soils can be performed by either using in-situ techniques or on-site extraction (following excavation). In the case of in-situ soil flooding, the aqueous extractive agent is allowed to percolate through the soil to promote metal mobilization. In the case of on-site extraction following excavation, the operation can be performed either batch-wise, semi-batchwise, or continuously. The soil is first pretreated for size reduction and classification. The contaminated soil is then brought into contact with the extractive agent, and the soil is separated from the spent extractive agent. The effluent is further recycled to decomplex

Table II. Properties of Chelating Agents

Chelating Agent	Molecular Weight	Acidity Constants					Metal Chelate log (Stability Constant)					
		pK_1	pK_2	pK_3	pK_4	pK_5	Cd	Zn	Cu	Pb	Mn^{++}	Fe^{+++}
NTA[a]	191	1.89	2.49	9.73	--	--	10.5	11.2	13.7	11.8	8.1	17.0
EDTA[b]	292	2.08	3.01	6.40	10.44	--	17.5	17.2	19.7	17.7	14.5	26.5
DTPA[c]	393	2.08	2.81	4.49	8.73	10.6	20.1	19.7	22.6	21.0	16.7	29.2

[a]Nitrilotriacetic acid

[b]Ethylenediaminetetraacetic acid

[c]Diethylenetriaminepentaacetic acid

and precipitate the metals from solution; alternatively, it is treated by using electrodeposition techniques to recover the metals.

Previous Studies Involving Extraction of Heavy Metals from Contaminated Soils. Ellis et al. (27) demonstrated the sequential treatment of soil contaminated with cadmium, chromium, copper, lead, and nickel using ethylenediaminetetraacetic acid (EDTA), hydroxylamine hydrochloride, and citrate buffer. The EDTA chelated and solubilized each metal to some degree; the hydroxylamine hydrochloride reduced the soil iron oxide-manganese oxide matrix, thereby releasing bound metals; it also reduced insoluble chromates to chromium(II) and chromium(III) forms. The citrate removed the reduced insoluble chromium and additional acid-labile metals. With the use of single shaker extractions, a 0.1 M solution of EDTA was much more effective for removing metals than was a 0.01 M solution. A pH of 6.0 was chosen as the optimal pH because it afforded slightly better chromium removal than that obtained at pH 7 or 8. The best single extracting agent for all metals was EDTA; however, hydroxylamine hydrochloride was more effective for removing chromium. Results of the two-agent sequential extractions indicated that EDTA was much more effective for removing metals than the weaker agents. The results of the three-agent sequential extraction showed that, compared with bulk untreated soil, this extraction removed nearly 100% of the lead and cadmium, 73% of the copper, 52% of the chromium, and 23% of the nickel. Overall, this technique was shown to be better than (1) three separate EDTA washes, (2) switching the order of EDTA and hydroxylamine hydrochloride treatment, and (3) simple water washes. When used alone, the EDTA washing can be effective, with only a slight decrease in overall removal efficiency. Lead was easily removed by the EDTA and was effectively removed by citrate. Cadmium was easily removed by EDTA and was effectively removed by the hydroxylamine hydrochloride; however, copper was only removed by the EDTA. Although nickel removal was poor with EDTA alone, the treatment with all three agents was no more effective.

Hessling et al. (15) investigated soil-washing techniques for the remediation of lead-contaminated soils at battery recycling facilities. Three wash solutions were studied for their efficacy in removing lead from these soils: (1) tap water alone at pH 7, (2) tap water plus an unspecified anionic surfactant (0.5% solution), and (3) tap water plus 3:1 molar ratio of EDTA to toxic metals at pH 7-8. Tap water alone did not appreciably dissolve the lead in the soil. Surfactants and chelating agents such as EDTA are potentially effective soil-washing additives for enhancing the removal of lead from soils. There was no apparent trend in soil or contaminant behavior related to Pb contamination (predominant Pb species), type of predominant clay in the soil, or particle size distribution. The authors concluded that the applicability of soil washing to soils at these types of sites must be determined on a case-by-case basis.

Elliott et al. (14) performed a series of batch experiments to evaluate the extractive decontamination of Pb-polluted soil using EDTA. They studied the effect of EDTA concentration, solution pH, and electrolyte addition on Pb solubilization from a battery-reclamation-site soil containing 21% Pb. The heavy metals concentrations in the soil were as follows: 211,300 mg Pb/kg (dry weight), 66,900 mg Fe/kg, 1383 mg Cu/kg, 332 mg Cd/kg, and 655 mg Zn/kg. A nine-step chemical fractionation scheme was used to speciate the soil Pb and Fe. Results from their study indicated that there was a positive correlation between increasing EDTA concentration and the release of Pb. (EDTA concentrations in the study ranged from 0.0004 to 0.08 M.) Recovery of Pb was generally greatest under acidic conditions and decreased modestly as the pH became more alkaline. Even in the absence of EDTA, a substantial increase in Pb recovery was observed below pH 5. As the pH became more alkaline, the ability of EDTA to enhance Pb solubility decreased, probably because hydrolysis was favored over complexation by EDTA. Elliott et al. observed that EDTA can extract virtually all of the non-detrital Pb, if at least a stoichiometric amount of EDTA is employed. When the amount of EDTA was increased above the stoichiometric requirement, it was capable of effecting even greater Pb recoveries. However, the Pb released with each incremental increase in EDTA concentration diminished as complete recovery was approached. The researchers also investigated the release of Fe from the soil by EDTA. The release of Fe increased markedly with decreasing pH. Although the total amount of iron was nearly 1.2 times the amount of lead in the soil, only 12% of the Fe was dissolved at pH 6 using 0.04 M EDTA, compared with nearly 86% dissolution of the Pb (12). Little of the Fe was brought into solution during the relatively short contact time of the experiments (5 h). The iron oxides retained less than 1% of the total soil Pb (12).

Elliott et al. (14) observed that Pb recovery increased by nearly 10% in the presence of $LiClO_4$, $NaClO_4$, and NH_4ClO_4. They attributed the increase to an enhanced displacement of Pb^{++} ions by the univalent cations and the greater solubility of Pb-containing phases with increased ionic strength. Below pH 6, calcium and magnesium salts also enhanced Pb recovery. Above pH 6, however, Pb recovery decreased because of competition between Ca or Mg and Pb for the EDTA coordination sites. Their research (12,14) did not provide any evidence that the suspension pH must be raised to at least 12 in order to prevent Fe interference in soil washing with EDTA for effective removal of Pb.

Goals and Objectives

The primary goals of this project were to determine and compare the performance of several chelating agents for their ability to extract lead from contaminated soil. The objective of this study was to determine the removal efficiencies of the various chelating agents (and water alone) as functions of solution pH and chelating agent concentration.

Experimental Procedure

The experiments were performed by using a batch shaker technique to investigate the extraction removal efficiency of lead from contaminated soil. The procedures followed in this study are summarized below.

Batch Shaker Test. Uncontaminated soil taken from a previous study was prepped by grinding the soil with a ceramic mortar and pestle so that the ground soil could pass through an 850-μm sieve (ASTM mesh #20). Characteristics of this soil are summarized in Table III. Four batches (~200 g each) of the uncontaminated soil were then spiked or artificially contaminated by soaking the soil with a solution of lead nitrate and left to dry. The high silt/clay content (~71%) results in a strong metal-soil affinity, favoring lead adsorption on the soil. Each batch of soil was "contaminated" with different lead nitrate concentrations (nominally 100, 500, 1000, 5000, and 10,000 mg/kg soil), which were assumed to be the initial Pb concentrations of the soil samples. Concentrations were checked by using nitric acid as an extractive agent and subsequent atomic absorption spectrophotometry (AAS) analysis. The contaminated soils were aged in glass jars (with screw-on covers) for approximately three months prior to the screening studies.

Soil was weighed by using a top-loading balance in 5-g portions and placed in plastic shaker containers that had lids. To these containers, 45 mL of one of the following solutions was added:

- 0.01 M, 0.05 M, 0.1 M, or 0.2 M Na_2-EDTA
- 0.01 M, 0.05 M, 0.1 M, or 0.2 M Na_3-NTA
- 0.01 M, 0.05 M, 0.1 M, or 0.2 M HCl
- Deionized water

Table III. Soil Analysis of the Uncontaminated Soil
Employed in Study

Parameter	Soil Analysis Noncontaminate
pH	7.75
Electrical Conductivity (mmhos/cm)	5.46
Cations (meq/L):	
Na^+	20.26
K^+	0.62
Ca^{++}	37.88
Mg^{++}	24.32
Anions (ppm):	
NO_3-N (1:5 H_2O extrac.)	2.7
P	10.3
Total Kjeldahl Nitrogen (TKN)	471
Fe (DTPA)	18.69
Sodium Absorption Ratio (SAR)	3.63
Calculated ESP	3.9
H_2O at saturation (%)	75.63
Cation Exchange Capacity (CEC), (meq/100 g)	13.27
Organic Matter (OM) (%)	2.02
Texture	Clay Loam
Size Fractions (%):	
Sand	29.2
Silt	35.0
Clay	35.8

These conditions were employed to create a matrix of samples; each combination of lead-spiked soil and type and concentration of chelating agent was tested. Replicate samples were prepared for every fourth sample. Contact times varied between 0.5 and 0.6 h.

The soil samples (to which the extractive agent had been added) were shaken for three hours at a low setting on an Eberbach shaker table. This time requirement was determined on the basis of results from preliminary experiments in which the residual concentration was monitored as a function of time on the batch shaker table. This analysis showed that chemical equilibrium could be achieved in three hours. Following agitation, the samples were centrifuged in plastic Nalgene centrifuge tubes equipped with snap-on caps, filtered by using No. 42 Whatman filter paper, and stored in glass vials maintained at pH < 2 (prepped using HNO_3) prior to AAS analysis. The Buck Scientific Atomic Absorption/Emission Spectrophotometer, Model 200-A, was calibrated by using AAS lead standards. The analyses were performed in accordance with the procedures described in **Standard Methods** (*28*).

Data collected during the experiments included the following: operating temperature, type of extractive agent and concentration, lead concentration on the soil (after treatment), lead concentration in the extractive agent after treatment, pH of solution before and after treatment, and batch shaking time.

Results and Discussion

Preliminary Experiments. Prior to performing the more comprehensive experiments involving the extraction of lead from the contaminated soil using EDTA and NTA at various concentrations of extractive agents and pH conditions, preliminary experiments were conducted using soil that had been contaminated with lead at 100 mg/kg soil. The purpose of these preliminary experiments was to determine the contact time required to reach pseudo-equilibrium conditions for a particular contaminated soil. The extractive agents studied included EDTA, NTA, and HCl (all at a concentration of 0.2 M). Figures 1 and 2 illustrate the results of these preliminary experiments by showing the lead solubilized into solution and the eluant pH (with no pH adjustment) as a function of time. The results show that pH quickly reached a constant value for each system; the pH corresponded to 4.2, 10.9, and 5.75 for HCl, NTA, and EDTA, respectively. Figure 2 indicates that the contact time required to reach pseudo-equilibrium conditions varies for each system. The approximate times for the HCl, NTA, and EDTA systems to reach pseudo-equilibrium conditions were 0.5, 3, and 2 h, respectively. To facilitate the treatment with chelating agents, a contact time of 3.0 h was selected to provide uniform treatment conditions. On the basis of the very low solubilization of lead when hydrochloric acid was used as the extractive agent, this system was not studied further.

Batch Extraction Experiments. A series of batch extractions was conducted in which water, NTA, and EDTA were used as the extractive agents. The initial lead concentrations on the soil were nominally 500, 1,000, 5,000, and 10,000 mg/kg soil after spiking the soil with the lead nitrate solution and allowing the samples to age at least six weeks. The concentrations of EDTA and NTA ranged from 0.01 to 0.10 M. The extractions were performed over the nominal pH range of 4 to 12.

Figure 3 presents the results for the extraction of lead using water alone for the 10,000 mg Pb/kg soil. The results for the other lead concentrations (ranging from 500 to 5,000 mg/kg soil) behaved similarly. The pH of the water was adjusted with the addition of HNO_3 or NaOH. The concentration of lead solubilized by the water extraction was pH-sensitive; a significant increase in lead concentration in solution is observed for pH < 4. Generally, the amount of lead solubilized by water alone is rather minimal; the concentration is generally below 10 mg/L. The maximum concentration observed in this study represented only 7.55% of the lead being solubilized.

Figure 3 also presents the results for the extraction of lead from the 10,000 mg/kg soil using 0.1 M NTA. The pH was adjusted with the addition of nitric acid or NaOH. Again, the results for the other lead concentrations behaved similarly. Whereas the lead concentration in solution using water as the extractive agent was as high as 81 mg/L, the lead concentration using NTA as the extractive agent was significantly higher, and concentrations as high as 212 mg/L were achieved at the lower pH range. The lead concentration in solution is observed to be much more pH-sensitive than the water system. Fairly low lead concentrations (indicating little lead solubilization) are observed over the pH range of 6 to 9. Significant increases in lead concentration are observed above and below this pH range. The maximum amount of lead solubilized over the pH range of 4 to 12 was 19.1%.

Figure 3 presents the analogous set of data for the extraction of lead from the 10,000 mg/kg soil using 0.1 M EDTA. Once again, the results for the other lead concentrations behaved similarly. In contrast to the other two extraction systems (employing water and NTA) that exhibited a significant pH dependency, the EDTA extraction system was strongly pH-insensitive. The lead concentration in solution ranged from 642 to 718 mg/L over the pH range of 4.9-11.3, corresponding to lead removals ranging from 59.5 to 64.2%. The pH insensitivity of the EDTA system can

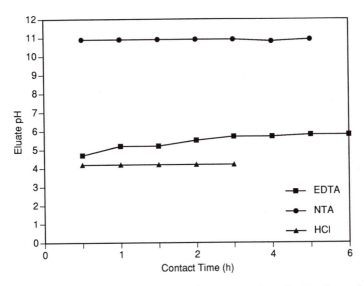

Figure 1. Eluate pH for extraction of lead from contaminated soil using various extractive agents.
Initial Pb Concentration = 100 mg/kg soil
Initial Extractant Concentration = 0.2 M

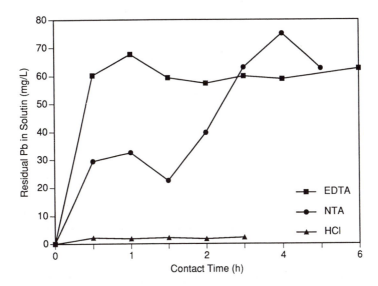

Figure 2. Lead concentration in solution following extraction of lead from contaminated soil using various extractive agents.
Initial Pb Concentration = 100 mg/kg soil
Initial Extractant Concentration = 0.2 M

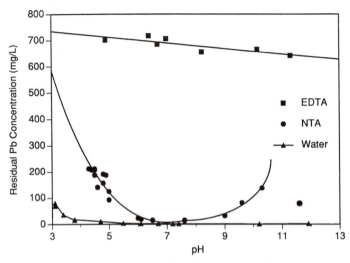

Figure 3. Comparison of the lead concentration in solution following extraction of
lead from contaminated soil for the extractive agents of water, NTA, and
EDTA
Initial Pb Concentration = 10,000 mg/kg soil
Concentration of EDTA and NTA = 0.1 M

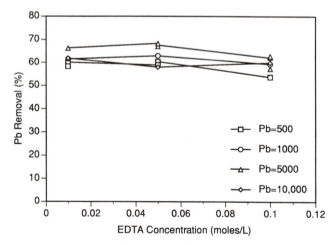

Figure 4. Comparison of the lead removal efficiency for EDTA extraction of lead as
a function of EDTA concentration.
Initial Lead Concentrations = 500, 1000, 5000, and 10,000
mg/kg soil

be explained by the stability constant of the EDTA-lead system, which is nearly 10^6 times larger than that for the NTA-lead system.

Figure 3 compares all three extractive agents in terms of the soluble lead concentration for removal of lead from the 10,000 mg/kg soil. The NTA shows the largest pH dependency, while the EDTA is pH-insensitive. The removal of lead remained relatively constant at approximately 61% over the entire pH range.

The effect of varying the EDTA concentration for the removal of lead from contaminated soil is shown in Figure 4. The removal efficiency is shown for four separate lead concentrations on the soil, ranging from 500 to 10,000 mg/kg soil. The removal efficiency was greatest for the case of a lead contamination level of 5000 mg/kg, although all the removal efficiencies were fairly constant, ranging from 54 to 68%.

Summary and Conclusions

The results from other research studies (*12,14*) indicate that removal of lead from contaminated soil can exceed 80%. The results of this study, which involved extraction of lead from a soil with a very high clay and silt content (35.8% and 35.0%, respectively) indicated that lead removal exceeds 60% using EDTA concentrations in the range of 0.01 to 0.10 M for lead concentrations on the soil ranging from 500 to 10,000 mg/kg. Extraction of lead using water alone removed a maximum of 7.55% for pH ~4. The removal of lead using water and NTA as extractive agents were both pH-dependent, whereas the removal of lead using EDTA was pH-insensitive. For the EDTA extraction system, the removal of lead ranged from 58 to 64% over the entire range of initial lead concentrations in the soil for the pH range of 4.9 to 11.3. The initial lead content had very little effect on the removal efficiency of lead for the EDTA system (for initial lead concentrations in the range of 500 to 10,000 mg/kg soil). The applied EDTA concentration over the range of 0.01 to 0.10 M also had little effect on the removal efficiency of lead from the soil.

Extraction of lead with EDTA was rapid, reaching equilibrium within a contact time of 1.0 h. Extraction of lead with NTA was slower; a contact time of approximately 3.0 h was required to reach equilibrium. The order of lead removal efficiency for the various extractive agents was as follows: EDTA >> NTA >> water. The maximum lead removals observed for this high clay and silt soil were 64.2, 19.1, and 6.8%, respectively, for the cases of EDTA, NTA, and water used as the extractive agents on soil with 10,000 mg Pb/kg.

Acknowledgments

The authors acknowledge the support of Argonne National Laboratory's Energy Systems Division, through the U.S. Department of Energy, under contract W-31-109-Eng-38, for funding of this research project. This paper was presented at the Spring National Meeting of the American Chemical Society held in Atlanta, GA, on April 14-19, 1991.

Nomenclature

AAS   Atomic absorption spectrophotometry

Cd   Cadmium

CEC   Cation exchange capacity, (meq/100-g)

Cu   Copper

DTPA	Diethylenetriaminepentaacetic acid
EDTA	Ethylenediaminetetraacetic acid
Fe	Iron
HCl	Hydrochloric acid
HNO_3	Nitric acid
K_i	Acidity constant for the ith dissociation
M	moles/L
Mn	Manganese
NTA	Nitrilotriacetic acid
OM	Organic matter
Pb	Lead
pH	$- \log [H^+]$
pK_i	$- \log (K_i)$
REDOX	Oxidation/reduction
SAR	Sodium absorption ratio
TKN	Total Kjeldahl nitrogen
Zn	Zinc

Literature Cited

1. Elliott, H.A.; Peters, R.W. "Decontamination of Metal-Polluted Soils Using Chelating Agents," In *Removal of Heavy Metals from Groundwaters;* CRC Press, Inc.: Boca Raton, FL, (in press).
2. Ku, Y.; Peters, R.W. "The Effect of Complexing Agents on the Precipitation and Removal of Copper and Nickel from Solution," *Particulate Science and Technol.,* **1988**, *6*, pp 441-466.
3. Ku, Y.; Peters, R.W. "The Effect of Weak Chelating Agents on the Removal of Heavy Metals by Precipitation Processes," *Environ. Prog.,* **1986**, *5*, pp 147-153.
4. Peters, R.W.; Ku, Y. "The Effect of Citrate, A Weak Complexing Agent, on the Removal of Heavy Metals by Sulfide Precipitation," In *Metals Speciation, Separation, and Recovery;* Patterson, J.W.; Passino, R., Eds.; Lewis Publishers: Chelsea, MI, 1987; pp 147-169.
5. Peters, R.W.; Ku, Y. "The Effect of Tartrate, A Weak Complexing Agent, on the Removal of Heavy Metals by Sulfide and Hydroxide Precipitation," *Particulate Science & Technol.,* **1988**, *6*, pp 421-439.

6. Oyler, J.A. "Remediation of Metals-Contaminated Site Near a Smelter Using Sludge/Fly Ash Amendments," In *Proc. 44th Purdue Indus. Waste Conf.*; Vol. 44; pp 75-82.
7. Hsieh, H.N.; Raghu, D.; Liskowitz, J.W.; Grow, J. "Soil Washing Techniques for Removal of Chromium Contaminants from Soil," In *Proc. 21st Mid-Atlantic Indus. Waste Conf., 1989, 21*, pp 651-660.
8. Trnovsky, M.; Oxer, J.P.; Rudy, R.J.; Hanchak, M.J.; Hartsfield, B. "Site Remediation of Heavy Metals Contaminated Soils and Groundwater at a Former Battery Reclamation Site in Florida," In *Hazardous Waste, Detection, Control, Treatment;* Abbou, R., Ed.; Elsevier Science Publishers, B.V.: Amsterdam, The Netherlands, 1988; pp 1581-1590.
9. Trnovsky, M.; Oxer, J.P.; Rudy, R.J.; Weinstein, G.L.; Hartsfield, B. "Site Remediation of Heavy Metals Contaminated Soils and Groundwater at a Former Battery Reclamation Site in Florida," In *Sixth Internat. Conf. on Heavy Metals in the Environ.;* S.E. Lindberg, S.E.; T.C. Hutchinson, T.C., Eds.; New Orleans, LA, 1987, Vol. I; pp 88-90.
10. Trnovsky, M.; Yaniga, P.M.; McIntosh, R.S. "Lead Removal — The Main Remedial Design Parameter for Two Florida Superfund Sites," In *Seventh Internat. Conf. on Heavy Metals in the Environ.;* Vernet, J.-P., Ed.; Geneva, Switzerland, 1989, Vol. II; pp 233-236.
11. Wagner, J.-F. "Heavy Metal Transfer and Retention Processes in Clay Rocks," In *Seventh Internat. Conf. on Heavy Metals in the Environment;* Vernet, J.P., Ed.; Geneva, Switzerland, 1989, Vol. I; pp 292-295.
12. Elliott, H.A.; Linn, J.H.; Shields, G.A. "Role of Fe in Extractive Decontamination of Pb-Polluted Soils," *Haz. Waste & Haz. Mater., 1989, 6,* pp 223-229.
13. *Electromembrane Process for Recovery of Lead Contaminated Soils: Phase I*; Final Report, National Science Foundation, Grant No. ISI-8560730, prepared by PEI Associates, 1986.
14. Elliott, H.A.; Brown, G.A.; Shields, G.A.; Lynn, J.H. "Restoration of Pb-Polluted Soils by EDTA Extraction," In *Seventh Internat. Conf. on Heavy Metals in the Environ.;* Vernet, J.-P.; Ed.; Geneva, Switzerland, 1989, Vol. II; pp 64-67.
15. Hessling, J.L.; Esposito, M.P.; Traver, R.P.; Snow, R.H. "Results of Bench-Scale Research Efforts to Wash Contaminated Soils at Battery-Recycling Facilities," In *Metals Speciation, Separation, and Recovery;* Patterson, J.W.; Passino, R., Eds.; Lewis Publishers, Inc.: Chelsea, MI, 1989, Vol. II; pp 497-514.
16. Wolfberg, G.A.; Kahanovich, Y; Avron, M.; Nissenbaum, A. "Movement of Heavy Metals into a Shallow Aquifer by Leakage from Sewage Oxidation Ponds," *Water Res.* **1980**, *14,* pp 675- .
17. Galbraith, J.H.; Williams, R.E.; Siems, P.L. "Migration and Leaching of Metals from Old Mine Tailings Deposits," *Ground Water* **1972**, *10,* pp 33- .
18. Jacks, G.; Arnemo, R.; Bagge, D.; Baerring, N.E. "Cleaning Up a Cadmium-Nickel Polluted Area;" In *Proc. Second Internat. Conf., Met. Bulletin Ltd.*; Worcester Park, England, 1980, pp 71- .
19. Winslow, G. "First Third Land Disposal Restrictions: Consequences to Generators," *Haz. Waste Manage. Mag.,* **1988**, Nov.-Dec.
20. Raghavan, R.; Coles, E; Dietz, D. *Cleaning Excavated Soil Using Extraction Agents: A State-of-the-Art Review;* EPA/600/2-89/034, U.S. Environ. Protection Agency, Risk Reduction Engineering Laboratory: Cincinnati, OH, 1989.
21. *Handbook for Remedial Action at Waste Disposal Sites;* EPA-625/6-82-006, U.S. Environ. Protection Agency, Municipal Environmental Research Laboratory: Cincinnati, OH, 1982.

22. Werther, J.; Hilligardt, R.; Kroning, H. "Sand from Dredge Sludge-Development of Processes for the Mechanical Treatment of Dredged Material," In *Contaminated Soil;* Assink, J.W.; van der Brink, W.J., Eds.; Martinus Nijhoff: Dordrecht, The Netherlands, 1986; pp 887- .
23. Aplan, F. "Flotation," In *Encyclopedia of Chemical Technology;* Grayson, M.; Eckroth, D., Eds.; Third edition; John Wiley & Sons: New York, NY, 1980, Vol. 10.
24. Pickering, W.F. "Metal Ion Speciation — Soils and Sediments (A Review)," *Ore Geology Reviews* **1986**, *1*, pp 83- .
25. Assink, J.W. "Extractive Methods for Soil Decontamination; A General Survey and Review of Operational Treatment Installations," In *Contaminated Soil;* Assink, J.W.; van der Brink, W.J., Eds.; Martinus Nijhoff: Dordrecht, The Netherlands, 1986; pp. 655- .
26. Slavek, J.; Pickering, W.F. "Extraction of Metal Ions Sorbed on Hydrous Oxides of Iron (III)," *Water, Air, and Soil Pollut.* **1986**, *28*, pp 151-162.
27. Ellis, W.D.; Fogg, T.R.; Tafuri, A.N. "Treatment of Soils Contaminated with Heavy Metals," In *Land Disposal, Remedial Action, Incineration and Treatment of Hazardous Waste*; 12th Ann. Res. Sympos., EPA 600/9-86/022, Cincinnati, OH, 1986; pp 201-207.
28. *Standard Methods for the Examination of Water and Wastewater*; 15th Ed., American Public Health Association: Washington, D.C., 1981.

RECEIVED February 8, 1992

Chapter 7

Surfactant Flooding
of Diesel-Fuel-Contaminated Soil

Robert W. Peters[1], Carlo D. Montemagno[2], Linda Shem[1],
and Barbara-Ann G. Lewis[3]

[1]Energy Systems Division and [2]Environmental Research Division, Argonne
National Laboratory, 9700 South Cass Avenue, Argonne, IL 60439
[3]Department of Civil Engineering, McCormick School of Engineering
and Applied Sciences, Northwestern University, Evanston, IL 60208

The use of surfactants to mobilize oil in water solutions has been
developed in the oil industry to enhance oil recovery from reservoirs. In
contrast, the use of surfactants as a soil-remediation technique is not as
well researched. This study used surfactant solutions to leach
undisturbed soil cores taken from a site that had been contaminated with
No. 2 diesel fuel after a puncture of a transfer line. Preliminary
screening of 22 surfactants was performed prior to this study to choose
the surfactants to be used in the flooding of the columns. The chosen
surfactants resulted in 87% to 97% removal of the diesel fuel during
batch extraction. Prior to flooding the columns with surfactants, the
columns were leached with deionized water to compare with leaching by
the surfactant. Water flooding resulted in removal of <1% of the diesel
fuel. Surfactant solutions removed generally <1%. The poor removal
efficiencies from the surfactant flooding, especially when compared
with results from the surfactant screening extractions, were attributed to
channeling of the solutions through the columns. The column
floodings, therefore, were essentially nonequilibrium extractions,
whereas the batch-screening extractions reached an equilibrium.

As our society becomes increasingly aware of the hazards associated with pollutants in
our soil and water supplies, it also becomes aware of the need for improved
remediation techniques. Remediation techniques such as pump and treat, excavate and
treat, or excavate and landfill are being implemented today with varying degrees of
success. The choice of technique depends on many factors, especially the location of
the contaminated area, hydrogeology of the site, characteristics of the soil, and nature
of the contaminant. This paper examines one experimental technique, in-situ surfactant
flooding, at the bench-scale treatability level, using undisturbed soil cores contaminated

0097–6156/92/0509–0085$06.00/0

with No. 2 diesel fuel. This technique was considered, along with air stripping and in-situ bioremediation, for a site in which approximately 60,000 gal of diesel fuel has spread to depths of up to 32 m (105 ft). The soil cores used in the bench-scale studies were taken from the contaminated site. Results from this study on surfactant flooding are presented and discussed.

Background

Surfactants. The use of surfactants in mobilizing petroleum oils from the ground is a well-known technique developed by the oil industry. Much research has been conducted on enhancing oil recovery from oil reservoirs by studying the interactions between surfactant solution, soil, and oil (1). The use of a surfactant as a valuable candidate for oil recovery involves its ability to solubilize oils in water. The fundamental properties of a surfactant responsible for this phenomenon are its amphipathic structure, monolayer orientation at interfaces, and adsorption at interfaces (2). Basically, the surfactant is made up of two functional groups, a hydrophilic head group and a lipophilic carbon chain. The two groups line up between the oil and water phases with their opposing ends dissolved in the respective phases. This arrangement creates a third layer at the interface, decreasing the interfacial tension (IT) between the oil and water. The balance of the head group and carbon tail determines which phase the surfactant molecule will dissolve into more readily. That is, if the head group is more heavily balanced, the surfactant as a whole will be more water soluble. In this case, it will tend to pull oil into solution as droplets encased in surfactant molecules. This balance is called the hydrophilic-lipophilic balance (HLB) (3).

The extent to which the IT is decreased depends largely on the concentration of surfactant in a particular solvent or liquid. Previous studies have shown that the point at which the IT is at or near its minimum is at the surfactant concentration where the most surfactant is found at the interface (4). This point is called the critical micelle concentration (CMC), the concentration at which aggregates or micelles begin to form from the excess surfactant in solution. Generally, the properties of a surfactant will vary linearly with increasing concentration until it reaches the CMC. At this point, the slope of the curve will change (5). Therefore, it is of interest to determine the CMC of a surfactant in a particular solution prior to its use.

Surfactants are categorized into four general groups, depending on the electrostatic charge of the head group. These groups are anionic, nonionic, cationic, and amphoteric. Anionic surfactants are widely used in oil recovery because they are highly water soluble and because their negative charge tends to be repulsed by a typical soil of a negative net surface charge. Head groups of important commercial anionic surfactants include sulfonates, sulfates, and phosphates. Nonionic surfactants are uncharged and soluble through hydrogen bonding at oxygen or hydrolyzed groups. Typical nonionic surfactants contain a polyoxyethylene group as the soluble group. Nonionics and blends of anionics/ nonionics have also been successful in their use as oil recovery enhancers. Cationic surfactants contain a positive functional head group, typically an amino or quaternary nitrogen group. Their use has been more as softeners and coating agents, and they do not usually perform well in soil of negative net surface charge (6).

Surfactants in Remediation. Although surfactant use in enhanced oil recovery has been explored in some detail, its use as a remedial flooding technique is still in the preliminary experimental stages, and few studies are available in the literature to date. Examples of some studies are discussed here.

Abdul et al. (7) studied the performance of commercially available anionic and nonionic surfactants to clean a sandy soil contaminated with automatic transmission fluid (ATF). Using a batch shaker method, Abdul et al. mixed 5 g of contaminated soil with 100 mL of 0.5% by volume surfactant solution for 30 min. After the soil settled, the decant and several rinses of the soil were analyzed for ATF. The soil was also analyzed for remaining ATF. Results showed that the surfactant removed between 56% and 84% of the contaminant but that washing the soil with water alone removed only 23%. Two of the better performers, removing 81% and 82%, were further screened at several different concentrations (0.1%, 0.25%, and 1.0%). For both of these surfactants, a concentration of 0.5% from the initial screening gave the best removal, while various other concentrations removed between 37% and 76%.

Ellis et al. (8) also tested several commercially available surfactants at the bench-scale level. They looked at treating both PCB(polychlorinated biphenyl)-contaminated soil and soil contaminated with hydrocarbons. The soil chosen was a freehold soil with the following characteristics: Total Organic Carbon (TOC), 0.12% (wt); cation exchange capacity (CEC), 8.6 meq/100 g; and compacted density, 1.68 g/cm.[3] The low CEC indicated an absence of mineralogic clay in the soil. Ellis et al. chose several surfactants and combinations of surfactants, including Richonate YLA (anionic) and Hyonic NP-90 (nonionic), which had been previously tested by the Texas Research Institute (TRI) to flush gasoline from sand. These surfactants were chosen on the basis of the following criteria: adequate water solubility, low clay-particle dispersion to avoid pore-space clogging, good oil dispersion, and adequate biodegradability to limit the environmental impact of the surfactant.

Ellis et al. then performed shaker tests similar to Abdul to screen the various solutions. They used 100 g of soil mixed with 200 mL of surfactant solution; the mixture was agitated for one hour. Again, both soil and decanted leachate were analyzed. A 1:1 blend of the nonionic surfactants Adsee 799 (Witco Chemical Co.) and Hyonic NP-90 (Diamond Shamrock) performed best, removing 93% of the hydrocarbons and 98% of the PCBs. These results were one to two orders of magnitude better than removal with water alone. Optimal concentrations of these surfactants were determined to be 0.75% by volume in further tests.

Following the screening process, Ellis et al. used the surfactant combination to perform flooding experiments on packed columns of Freehold soil spiked with PCBs and chlorinated phenols. Initially, columns were flooded with water. This treatment was followed by several pore volumes of the surfactant solution and then a final wash of water. The percolation rate was 1.5×10^{-3} cm/s at a 60-cm constant head. Results showed the initial and final water rinses to be ineffective in reducing contamination. However, after three pore volumes of surfactant solution, 74.5% of the pollutant was removed. After 10 pore volumes, 85.9% removal resulted. Further studies on column flooding using various concentrations of the surfactant combination verified the optimum of 0.75% by volume obtained in the screening process.

Several studies are taking bench-scale results into the field at the pilot-scale level. Following screening and column studies performed in the laboratory, Nash (9) tested some of the more successful surfactants on a site contaminated with waste solvents, waste lubricating oil, and JP-4 fuel. In 10 holes dug 1 ft deep, Nash applied solutions at 3 in. per day for four to six days. Conclusive results were not obtained because of inherent difficulties in working in the field. Nash encountered two major problems: (1) difficulty in obtaining accurate contamination levels from sampling because of the nonuniform distribution of the contaminant on the soil and (2) dilution of the flooding solution because of rainy weather during the surfactant application period.

Surfactant flooding as a technique still has far to go before it will be considered a demonstrated and proven technology. Because of the uniqueness of each site and the interaction between surfactant, soil, and contaminant, even pilot-scale studies must be fine-tuned before conclusive results can become available for the development and application of this technique. This paper contributes to the development of this technique by presenting results of column-flooding tests on undisturbed soil cores.

Methods

The soil used in this study was taken from a site in California that had been contaminated with No. 2 diesel fuel. Undisturbed soil cores were obtained by using a split spoon auger from one boring reaching to a depth of 32 m (105 ft) and cut into 38-cm (15-in.) columns with a diameter of 18 cm. Ten of these columns were cut into halves of 19 cm (7.5 in.). The top halves were used for the leaching and flooding experiments.

Characterization of the soil was performed on soil samples taken approximately 2 in. from the surface of each column separately. Tests and methods employed are listed in Table I.

The columns were first leached with three effective pore volumes (EPV) of deionized water and then drained. EPVs were estimated by noting the time it took for initial flooding of the entire soil core at a constant flow rate. Calculated pore volumes for the columns were approximately three times greater than the estimated effective pore volumes measured. Flooding the columns for three to four EPVs with a 1.5% surfactant solution followed leaching them with water. Leaching and flooding of the columns were performed using a constant head, up-flow method through aluminum bases and tops. The original core polybutyrate casings served as column walls so as not to disturb the soil samples.

Preliminary screening of several commercially available surfactants was performed with the soil using a batch-shaker test. Three of the better performers from this screening were chosen for the subsequent flooding experiments. The initial plan was to flood at least nine columns, thus allowing triplicates of each of the three chosen surfactants. Because of these constraints and disturbances of the columns (especially the clogging of column chamber inlets while draining the columns after leaching), only four columns were flooded. The columns chosen and the surfactants

Table I. Soil Characterization Tests
Performed on Grab Samples Taken
from Surface of Each Soil Core

Tests	Methods[a]
Hydraulic Conductivity[b]	EPA 9100
pH	EPA 9045
Cation Exchange Capacity	EPA 9081
Porosity	MSA
Soil Dry Weight	MSA
Particle Size, Dry Sieve	ASTM
Particle Size, Hydrometer	MSA

[a]Reference Sources: EPA-"Test Methods of Evaluating Solid Waste" (10), MSA-"Methods of Soil Analysis, Part 1" (11), and ASTM (12).

[b]Determined on 2-in.-diameter, 2-in.-long cores taken from the top of the soil column.

used for each column are listed in Table II. Characteristics of the soil in each column are summarized in Table III. Results of soils collected at other depths are also presented in Table III for comparison.

Each EPV of surfactant eluate and drained leachate was collected separately. Each collection was subjected to liquid-liquid extraction using a carbon disulfide solvent at a 0.025:1.0 solvent/solution ratio and analyzed for total petroleum hydrocarbons (TPH) by gas chromatography (GC). Soil grab samples taken from the top of each column before and after flooding were subjected to a slurry extraction at a 1.0:1.0 solvent/solution ratio also using carbon disulfide. Then, they were analyzed by GC for TPH. The extraction method was modified from the "California Leaking Underground Fuel Tank Manual" (*13*) and Wilson (*14*). After all treatments were completed, the columns were destroyed, the soils from each column were mixed, and duplicate grab samples from each of the mixed soils were analyzed for TPH. Alkanes between C12 and C19, key components of the diesel fuel, were identified and their concentrations were determined by GC to indicate the presence of selective mobilization by the surfactant solution.

Table II. Soil Columns and Surfactants Used in
Bench-Scale Surfactant Flooding Tests

Column Depth (m)	Surfactant/ Code Number	Surfactant Type	Chemical Name
6.3	PSVS/18	Anionic	Polysodium Vinyl Sulfonate
17.1	Cyanamer P-35(L)/15	Anionic	Sodium Polyacrylate
23.2	Cyanamer P-35(L)/15	Anionic	Sodium Polyacrylate
26.2	Surfynol 485/13	Nonionic	Exthoxylated Tetramethyl decynediol

Table III. Results of Soil Characterization Tests Performed on Samples Taken from
Soil Cores Employed for Leaching and Flooding Experiments

Depth (m)	Hydraulic Conductivity (cm/s)	Porosity[a] (%)	pH	CEC[b] (meq/100-g)	Dry Soil (%)	Texture Class
6.3	7.03×10^{-4}	45.34	8.3	16.03	88.98	Sandy Loam
8.4	4.23×10^{-4}	42.84	8.1	23.98	86.9	Loam
14.8	2.39×10^{-3}	36.87	7.7	19.58	86.0	Sandy Loam
17.1	1.64×10^{-3}	28.95	7.5	19.67	87.92	Sandy Loam
23.2	5.43×10^{-5}	26.2	7.9	14.51	87.6	Clay Loam
26.2	1.15×10^{-4}	45.81	7.7	8.88	89.05	Sandy Loam
32.0	6.79×10^{-3}	42.64	7.8	19.39	86.99	Loam

[a]Porosity was measured from samples taken off the first 2 in. of the core.

[b]Cation Exchange Capacity.

Results

Surfactant Screening. Partial results from the surfactant screening tests are presented in Table IV. (For complete results see Peters et al. [15]) Surfactants 13 (nonionic), 15, and 18 (both anionic) were chosen for column-flooding studies on the basis of their high level of performance in removing TPH from soil during the screening process.

Column Flooding. Table V lists results of the TPH analysis taken from soil grab samples both before and after flooding as well as the percent removal as indicated by these values. A wide range of percent removal, between -42% and 68%, resulted. A minus percent removal indicates more TPH was measured on the treated soil compared with that measured before treatment.

Table VI compares TPH values from the soil grab samples taken from the tops of the columns at different times during the water leaching and surfactant flooding

Table IV. Partial Results from Batch Extraction
Surfactant Screening Tests[a]

Surfactant (Code No.)	% Removed	% Enhanced[b]	Surfactant Type
Surfynol 485 (13)	97.90	45.09	Nonionic
Surfynol 468 (14)	95.84	42.03	Nonionic
Cyanamer P-35(L) (15)	93.00	37.84	Anionic
PSVS (18)	87.28	29.35	Anionic
Cyanamer P--70 (17)	80.00	18.56	Anionic
Emcol CC-42 (1)	65.68	negli.	Cationic
Witconol 2648 (6)	9.55	negli.	Nonionic
Control - plain water	67.48	---	---

[a]Ratio of soil to surfactant was 10 mg soil to 50 mL of 2% surfactant solution. Shaking time was 1 h.

[b]Percent of TPH removed by the surfactant relative to the percent of TPH removed by water alone.

Table V. Total Petroleum Hydrocarbons, TPH, in Soil
before and after Flooding Measured by
Grab Sampling on Intact Columns

Depth (m)	Total Petroleum Hydrocarbons (mg/kg)		Removal (%)
	Before Flooding	After Flooding	
6.3	1236	1024	17.15
17.1	33	46.9	-42.09
23.2	328	225	31.4
26.2	18132	5823	67.89

Table VI. Total Petroleum Hydrocarbons, TPH, on Soil Taken from
Column Tops at Various Stages of Leaching and Flooding
and from Mixed Column Soil after Flooding

Column Depth (m)	Total Petroleum Hydrocarbons, (mg/kg				TPH from Mixed Column Soil (mg/kg)
	Before Leaching	After H$_2$O Leaching	Before Surfactant Flooding	After Surfactant Flooding	
6.3-a	890	1060	930	1020	1460
6.3-b			1540		1280
8.4-a	3650	7210	7520	-c-	3590
8.4-b					8360
17.1-a	4180				2780
17.1-b	4570	33	33	46.9	3590
23.2-a					777
23.2-b	74.5	328	330	225	874
26.2-a					10800
26.2-b	795	4660	18100	˙ 5820	11100

Note: a and b indicate duplicate samples; c denotes that this column was not
subjected to surfactant flooding experiments.

experiments with samples taken from the well-mixed column soil after columns were broken up following water leaching and surfactant flooding. No apparent decrease was observed in the mixed soil following treatment, and the TPH values were higher in some cases. Duplicates taken on soil grab samples varied by orders of magnitude in some cases. Duplicates from the well-mixed soil generally differed by 15% or less.

Table VII compares the TPH concentration (in mg/kg soil) measured in the total leachate collected from the floodings with TPH on the soil before flooding. The percent removal for the flooding was very low (<1%) with the exception of one core, the 17.1-m column, which showed a removal of slightly >9%.

Figure 1 represents the TPH measurements in mg/L for each EPV collected by leaching the columns with deionized water and subsequently flooding them with surfactant solution. The dotted line in the middle of each curve indicates points between the final water leachate collection and the first surfactant flooding collection. Generally, a downward trend is seen with each successive collection of the water leachate, with the exception of the 23.2-m column. For the surfactant flooding, the 17.1-m and 26.2-m columns showed an increase in TPH removal as more EPVs of solution was passed through the column. The 6.3-m column showed an improvement in TPH removal for the first surfactant collection over that of the water leachate. No more TPH was removed with further EPVs addition, however. The 23.2-m column also showed a general drop in TPH removal with surfactant flooding, except for the last collection. A break in the curve between the seventh and ninth EPV for the 26.2-m column indicates where the column had been drained and stored for one month before being resaturated and flushed with two additional pore volumes. The columns were also drained between leaching and flooding.

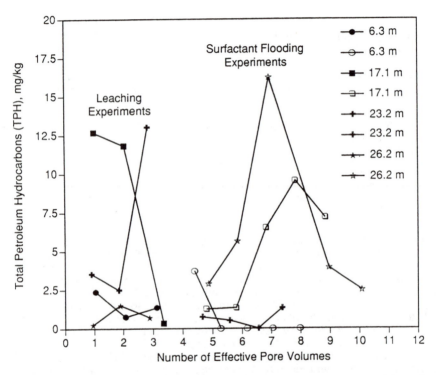

Figure 1. Total petroleum hydrocarbon (TPH) concentrations measured in each leachate collected for leaching and surfactant flooding experiments.

Table VII. Total Petroleum Hydrocarbons, TPH, Measured on Soil
Before Flooding and in Leachate from Surfactant Flooding

| Column Depth (m) | Total Petroleum Hydrocarbons (mg/kg) | | % Removal |
	Before Flooding	In Leachate from Surfactant Flooding	
6.3	1240	0.68	0.055
17.1	33	3.05	9.242
23.2	328	0.33	0.101
26.2	18100	3.79	0.021

Table VIII compares the amounts (mg) of TPH removed in approximately three EPVs of leachate with water with approximately four EPVs after flooding with surfactant. The 6.3-m column using surfactant 18 (S-18) showed similar amounts removed with and without surfactant. The 17.1-m column using S-15 also resulted in similar removals between the two leachates, with a slight improvement for the case of surfactant application. The 23.2-m column using S-15 showed a drastic decrease, whereas the 26.2-m column using S-13 indicated a substantial increase.

Figure 2 shows removal of individual alkanes in the flooding eluate. No removal of alkanes occurred for the 6.3-m column using S-18. The 17.1-m column showed increased removal of alkanes (as it did for TPH) with each successive flooding, peaking with the fourth EPV. The concentrations of C14 and C15 were as high as they were initially on the soil. The 23.2-m column had little removal of alkanes (as with TPH), C14 and C15 generally having the greatest concentration of the alkanes again. The 26.2-m column, however, seemed to favor C13 and C17 for the first two collections. No alkanes were found in the collections of the fourth and fifth EPV or for the second drain of the 26.2-m column. These were collected after the column had been drained and in contact with the surfactant for more than a month.

Table VIII. Total Amounts of Total Petroleum
Hydrocarbons in Water and Surfactant
Leachates Successively

| Column Depth (m) | Total Petroleum Hydrocarbons (mg) | |
	Deionized Water (~ 3 Pore Volumes)	Surfactant Leachate (~ 4 Pore Volumes)
6.3	6.68	4.46
17.1	21.83	24.09
23.2	16.63	2.35
26.2	1.85	26.66

Discussion

The percentage of TPH leached from the floodings varied over a wide range of values. These final percentages were based on initial concentrations measured on the soil (Table V). The initial values were obtained by extracting TPH from samples taken off the top

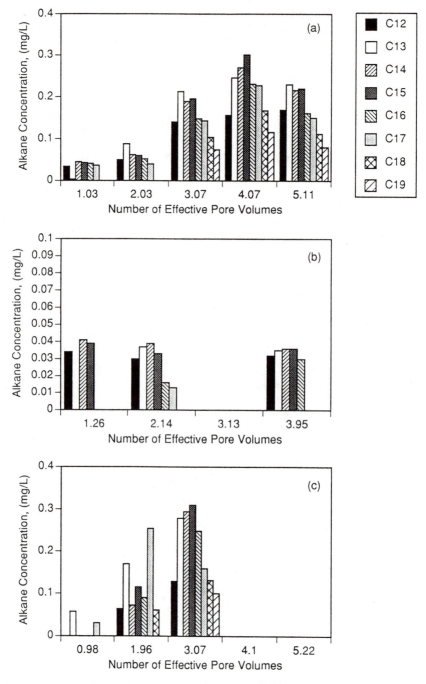

Figure 2. Individual alkane concentrations measured in each effective pore volume collected during surfactant flooding for columns (a) 17.1-m, (b) 23.2-m, and (c) 26.2-m.

surface of the column (i.e., grab samples) and were assumed to represent the entire column. Because of the presumed nonuniform distribution of the contaminant on the soil and nonuniform migration of the contaminant upward through the column while flushing, obtaining a good representative value was not possible. Variation in TPH values of the grab samples was apparent when comparing these values with those obtained from the well-mixed soil after the columns were destroyed (see Table VI). The differences in the values before and after flooding (see Table V) were generally within one order of magnitude. Statistically, these differences may not be significant. (Multiple sampling for statistical evaluation was not done in order to prevent core disturbance.) If these values are not significantly different, then Table V indicates that no measurable decrease in contamination levels resulted from the surfactant floodings.

A better representation of TPH removed from the core is obtained by comparing TPH values of the total leachate to initial TPH values on soil (Table VII). In this case, extraction was performed on the total volume of leachate rather than a grab sample involving the soil. Also, the problem of nonuniform distribution that is inherent with soils is not a major consideration in liquids. Additionally, the values obtained with this method were more consistent than those obtained with soil grab samples (compare Tables 5 and 7).

A comparison of TPH removal values obtained by leaching columns with water alone and with surfactant solution (Table VIII) shows that the only substantial increase in amounts of TPH removed was observed in the 26.2-m column (S-13); a dramatic decrease was observed in the 23.2-m column, and only slight differences were observed in the 6.3-m and 17.1-m columns. These results indicate that the use of surfactants 15 and 18 did not yield an improvement over using water alone, but surfactant 13 was quite successful in enhancing the diesel fuel mobilization. The generally poor performance of these surfactants is in contrast to the screening results (Table IV) and the results of other studies (6-9).

An alternative explanation for the low removals in the column tests when compared with the screening test (15) involves laboratory observations. Extensive channeling was observed in the columns after they were handled, and they were subject to some moisture loss before the leaching and flooding studies were performed. The leachate, therefore, likely passed through the columns with relatively little contact with the soil. Residual oil in the soil bulk matrix was most likely unaffected by these floodings. Some TPH concentrations in the water leaching were relatively higher than those for the surfactant floodings possibly because of water initially displacing diesel fuel from large pores in a plug-flow manner.

Because of the artifact introduced by channeling, leachate volumes are expressed as effective pore volumes, defined as the volume of water necessary to initially flood the soil columns. Figure 1 shows the concentration of TPH measured in each effective pore volume collected for both water leaching and surfactant flooding. Initially the following trends (assuming uniform soil/solution contact) were predicted for most of the columns: (1) an initially high contaminant concentration in the first leaching that would have represented the oil front because of piston flow and (2) with the addition of the surfactant solution, an increase in TPH removed with each successive flushing until an equilibrium plateau would be reached. The trend for the four columns in the water leaching portion of the collections did show a general decrease with subsequent pore volumes except for the 17.1-m column. This most likely indicated that the initial passes of water through the column displaced the oil in the larger pores. What was observed for the surfactant flooding collections, however, turned out to be different for each column. The 6.3-m column using S-18 indicated the presence of TPH only during the first surfactant collection. The 23.2-m column (S-15) also showed very low concentrations (even lower those that seen in the water leachates), which decreased with each collection. The 17.1-m columns using S-15 and the 26.2-m column using S-13, however, both showed increasing TPH concentrations during the first few collections. For the 26.2-m column, a plateau may have been reached if the flooding had not been interrupted. The 17.1-m column shows a peak concentration in leachate for the fourth pore volume. This peak may have been caused by the surfactant causing dispersion of colloidal-sized particles into solution. Surfactant 15 may have caused eventual clogging of soil pores by surfactant/soil or surfactant/oil aggregates, resulting in a decrease of contaminant in the leachate. Surfactant 18 may also have reacted with the soil in this way. The potential needs to be examined for these surfactants to cause soil dispersion and clogging of pores.

Because each column was unique, other variables in addition to channeling likely played key roles in the results obtained. These variables include differences in solution/soil contact, initial concentration of TPH in the core, nonuniform distribution of TPH, variation in soil texture, and dispersion of soil aggregates by the particular surfactant.

Alkanes C12-C19 were shown by GC to be the main constituents of the diesel fuel found in the soil cores. The concentrations of each compound in the successive effective pore volumes were compared to determine if preferential mobilization was occurring. Before leaching, C14-C16 in the soil were the most prevalent of the alkanes observed; C13, C17, and C18 were substantially lower; and C12 and C19 had the lowest concentrations. This pattern was generally found in the leachates for the 17.1-m column using S-15 and the 23.2-m column using S-15, indicating no preferential mobilization of these alkanes (Figure 2). The first two pore volumes collected from the 26.2-m column using S-13 showed, however, possible preferential mobilization of C13 and C17. This pattern is not repeated in subsequent collections, suggesting that the three surfactants caused no preferential mobilization of any one of these alkanes.

Although these results may have been the result of unavoidable experimental disturbances for the undisturbed cores collected from the site, they are a good indication of what may happen in the field when implementing this technique in a saturated soil, especially in soils that have low hydraulic conductivity of the bulk matrix, lenses of clay or sand, and fractures. The flooding fluids will take the highest-pressure-gradient path through sand lenses or fractures and fail to contact the bulk of the soil matrix where much of the contaminant may be contained if the spill occurred prior to fracturing. After the initial spill, however, the diesel fuel can be expected to redistribute, especially after a number of years, in such a way that subsequent transport of the contaminant will occur primarily in the bulk matrix of finer pores rather than in the fractures. Travis and Doty (16) have discussed problems with on in-situ flooding or pump and treat methods for remediating contaminated aquifers. They point out that none of the numerous superfund sites that have implemented a pump and treat remediation technique have seen a substantial lowering of contaminant levels even after 10 years of continuous treatment. They indicate reasons for this behavior include sorption of water insoluble contaminants to organic matter in soils as well as the preferential flow of water through high permeability channels at the site. Although the use of a surfactant solution will aid in mobilizing insoluble contaminants, as seen by the success of the equilibrium-extraction batch-screening process performed prior to this study, the problem of preferential flow in cracks seems to dominate success of contaminant removal by flooding. The use of surfactants to decontaminate soil may be better employed using an on-site soil-washing technique following excavation.

Summary and Conclusions

Use of surfactants may prove to be a good technique for separating diesel fuel from a soil as indicated in results obtained in preliminary screening tests (removal efficiencies up to 97%) and from results of other studies performed on laboratory-packed soil columns (removal efficiencies up to 8.6%), assuming flow through the bulk matrix. Results presented here using undisturbed, diesel-fuel-contaminated soil cores taken from a site indicate, however, that removal of diesel fuel from the soil flooded with surfactant solution was generally less than 1%. Low removal efficiencies in these soil cores are attributed to experimental artifacts such as channeling of the surfactant solution through cracks developed unavoidably in the cores during shipping, handling, and leaching.

As an applied technique, these results suggest that in-situ surfactant flooding, especially for sites that contain fractures or are prone to channeling, is not an effective method of removing diesel fuel from the soil. The use of a surfactant solution for washing the soil on-site after excavation could, however, be an effective soil-remediation technique, given the right surfactant or combination of surfactants for the site and contaminant.

Nomenclature

ASTM   American Society of Testing and Materials

ATF   automatic transmission fluid

CEC   cation exchange capacity, meq/100-g

CMC   critical micelle concentration

EPA   U.S. Environmental Protection Agency

EPV   effective pore volume

GC   gas chromatograph

HLB   hydrophilic-lipophilic balance

IT   interfacial tension

MSA   Methods of Soil Analysis

pH   - log $[H^+]$

PCB   polychlorinated biphenyl

PSVS   polysodium vinyl sulfonate

TPH   total petroleum hydrocarbons

TRI   Texas Research Institute

Acknowledgments

This work was supported by the U.S. Department of Energy, Albuquerque Operations Office, under contracts W-7405-Eng-36 and W-31-109-Eng-38. This paper was presented at the Spring National Meeting of the American Chemical Society, held in Atlanta, GA, on April 14-19, 1991.

Literature Cited

1.   *Tertiary Oil Recovery Processes Research at the University of Texas: Final Report*; DOE/BC/20001-11, Prepared by Texas Research Institute for the U.S. Department of Energy, April 1983.

2.   "Surfactants," In *Kirk-Othmer Encyclopedia of Chemical Technology, 2nd ed.*;
     Dukes, E.P.; Henson, E.P.; Hutto, N.; Klingsbery, A.; Maxwell, J., Assoc.
     Eds.; John Wiley and Sons, Inc.: New York, NY, 1968: pp 507-593.
3.   *Using Oil Spill Dispersants on the Sea*; National Research Council, National
     Academic Press: Washington, D.C., 1989.
4.   *Developments in Petroleum Science 17B, Enhanced Oil Recovery, II: Processes
     and Operations*; Donaldson, E.C.; Chilingarian, G.V.; Yen, T.F., Eds.; Elsevier
     Science Publishing Company, Inc.: New York, NY, 1989.
5.   *Kirk-Othmer Encyclopedia of Chemical Technology*; Standen, A., Ed.; Vol. 19;
     John Wiley and Sons: New York, NY, 1969, pp 508-593.
6.   Richabaugh, J.; Clement, S.; Lewis, R.F.; "Surfactant Scrubbing of Hazardous
     Chemicals from Soil," In *Proc. 41st Purdue Indus. Waste Conf.*; 1986, pp. 377-
     382.
7.   Abdul, A.S.; Gibson, T.L.; Rai, D.N.; "Selection of Surfactants for the Removal
     of Petroleum Products from Shallow Sandy Aquifers," *Groundwater* **1990**, *28*,
     pp 920-926.
8.   Ellis, W.D.; Payne, J.R.; McNabb, G.D.; *Treatment of Contaminated Soils with
     Aqueous Surfactants*; EPA/600/2-85/129, U.S. Environmental Protection
     Agency, Hazardous Waste Engineering Research Laboratory: Edison, NJ,
     November 1985.
9.   Nash, J.H. *Field Studies of In-Situ Soil Washing*; EPA/600/2-87/110, U.S.
     Environmental Protection Agency: Washington, D.C., December 1987.
10. Black, C.A.; Evans, D.D.; White, J.L.; Ensminger, L.E.; Clark, F.E., Eds.;
     *Methods of Soil Analysis: Part 1 Physical and Mineralogical Properties,
     Including Statistics of Measurement and Sampling*; American Society of
     Agronomy, Inc.: Madison, WI, 1966.
11. *Test Methods for Evaluating Solid Waste Volume IC: Laboratory Manual,
     Physical/Chemical Methods*; EPA/SW-846, U.S. Environmental Protection
     Agency: Washington, D.C., November 1986.
12. *Annual Book of ASTM Standards, vol. 4.08: Soil and Rock; Building Stones;
     Geotextiles*; D422-63, American Society of Testing and Materials: Philadelphia,
     PA, 1990.
13. Leaking Underground Fuel Tank Task Force, *Leaking Underground Fuel Tank
     Field Manual: Guidelines for Site Assessment, Cleanup, and Underground
     Storage Tank Closure*; No. 21224473, State Water Resources Control Board,
     State of California: Sacramento, CA, October 1989.
14. Wilson, J.T.; Leach, L.E. *In Situ Bioremediation of Spills from Underground
     Storage Tanks: New Approaches for Site Characterization, Project Design, and
     Evaluation of Performance*; EPA/600/2-89/042, U.S. Environmental Protection
     Agency: Washington, D.C., July 1989.
15. Peters, R.W.; Montemagno, C.D.; Shem, L.; Lewis, B.G. "Surfactant Screening
     of Diesel-Contaminated Soil," In *Proc. Third Annual IGT Internat. Symp. on
     Gas, Oil, Coal, and Environmental Biotechnology*; New Orleans, LA, December
     3-5, 1990.
16. Travis, C.C.; Doty, C.B. "Can Contaminated Aquifers at Superfund Sites be
     Remediated?," *Environ. Sci. & Technol.* **1990**, *24*, pp 1464-1466.

RECEIVED February 3, 1992

Chapter 8

Separation Steps in Polymer Recycling Processes

Ferdinand Rodriguez, Leland M. Vane, John J. Schlueter, and Peter Clark

School of Chemical Engineering, Olin Hall, Cornell University,
Ithaca, NY 14853

The recycling of polymers is generally considered to be the most environment-friendly alternative for dealing with the disposal problem. Even if consumers can be persuaded to segregate polymer products before putting them into the municipal waste stream, there will always be the matter of separating (a) inadvertent mixtures and (b) multicomponent products. One process which seems economically attractive in the current market is to separate "impurities" from a stream which is predominantly one polymer by using differences in density. The poly(ethylene terephthalate) beverage bottle with a polyethylene base is a case in point. Of course, there is the alternative of not separating the components of a waste stream, but to make a "plastic lumber" out of a mechanical mixture of polymers. A greater degree of separation and <u>purification</u> is possible if the polymers are put in solution or are converted to small molecules. Multicomponent plastic products can be treated if suitable conditions are found for selective dissolution.

To what extent must a polymer be purified for re-use? Almost any mixture of polymers will be multiphased since few polymers are mutually compatible even when they have chemically similar structures. Optical transparency is invariably lost in a mixture and strength usually suffers. A separation step is invariably a benefit for the re-use of a waste stream. The more homogeneous the product, the broader its field of applications.

The degree to which separation is feasible and the degree to which it is necessary most often are dictated by economics although legislative considerations may apply. In some communities, for example, curbside separation of paper, glass, metal, and plastics from other garbage and trash may be mandated by law. We will restrict ourselves here to dealing with a polymer waste stream ostensibly free of most non-polymeric material.

The extension of life for a polymer in a "down-graded" application may tolerate a wide range of mixtures. "Plastic lumber" can be made from various mixtures where the application (thick-walled containers, park benches) can compensate for low

0097–6156/92/0509–0099$06.00/0
© 1992 American Chemical Society

strength by additional total amount of material. Since any extended life of this sort does not decrease the rate of discard in the steady state, such applications are not recycling in the sense of decreasing the net market input or waste output streams.

True recycling on a "global" basis demands that some part of the total polymer market input stream be replaced by reclaimed material. This, in turn, requires a purification of contaminated waste polymer to a state that allows it to replace material in the same application or, at least, to replace the polymer in another existing application currently using "virgin" material.

Alternatives to Recycling. There are alternatives to recycling [1,2]. Burial in "sanitary landfills" would probably not be objectionable if the input were clean rather than contaminated with garbage. However, the public would not be likely to countenance spending government funds on cleaning polymers followed by unprofitable interment. Photo- and bio-degradable polymers represent answers to the aesthetic problem of litter and can be regarded in some respects as forms of oxidation without energy recovery[3,4]. Incineration [5], especially with energy recovery, can be more easily justified. As to the arguments that resources are wasted or that carbon dioxide emissions are augmented, one can answer with the following example. The 5 to 8 gram high density polyethylene grocery bag offered in a supermarket may be declined by an environmentally-aware customer. However, before the customer has driven an automobile out of the parking lot, he or she will have burned hydrocarbons (a mixture of linear, cyclic and aromatic compounds) amounting to several times the weight of the bag. In fact, all polymers combined (about 75 billion pounds/year) amount to less than 3% of the petroleum and natural gas products used as fuels in the United States (over 2,500 billion pounds/year excluding coal). The separation steps required for use of mixed polymers as fuels in central incinerators are minimal or not needed in most cases. Moreover, the control over emissions can be made much more effective than in distributed sites such as homes or automobiles.

Another alternative to recycling is pyrolysis [6]. While the idea of material resource recovery is attractive, a mixed-feed stock generally yields a pyrolysate of complex nature. One has but to remember the coal pyrolysis attempts of a generation ago to appreciate the difficulty of making effective use of a mixture of hundreds of chemical intermediates.

For some polymers, depolymerization is possible [1]. Poly(ethylene terephthalate), PET, can be treated with methanol to reform dimethyl terephthalate and ethylene glycol, both of which can be refined by distillation and reused to make new polymer. Polystyrene and several other less common polymers form yields of monomer on pyrolysis [7]. Generally, a rather homogeneous source of polymer is needed as a feed for effective depolymerization. Most commercial examples appear to have used industrial rather than consumer waste. PET can be changed by other transesterification or hydrolysis reactions to various uses. So, to some extent, can polyurethanes. Somewhat similar is the devulcanization of rubber in which a sulfur-crosslinked network is converted to a system of more-or-less thermoplastic components [8]. Rubber reclaiming has fallen on hard times for various reasons including the difficulty of making an environmentally attractive process.

Experience with aluminum, glass and paper. There are lessons polymer recyclers can learn from older programs. Recycling of aluminum beverage cans is a true success. Over 50% of all cans were recycled in 1988, mostly back to the original application [9]. Several factors leading to this situation are worth considering for polymers.

a.   The cans are all aluminum. Older cans which had some parts of ferrous alloys were not easily recycled.

b. All vendors use the same material. If one used a modified alloy that was not compatible with that used by the others, the cans would have to be segregated before recycling.

c. The value of scrap aluminum is high (c. $0.70/lb).

On the other hand, only about 28% of aluminum goes into cans. Wire, foil, and appliances are not as easily recycled. Also, most non-beverage metal containers (foods, personal care products, etc.) are not aluminum. Almost 29% of the metal cans used in 1987 were steel (both tin-plated and tin-free) [10]. Not since World War II days has there been a booming market for "tin can" scrap.

About 8.5% of glass containers used in 1986 were recycled [11]. Since only about 20% of all silica-based glass goes into other products (flat glass, fiberglass, etc.), the 8.5% figure is rather impressive. Most recyclers request (and most municipalities that recycle require) that glass be separated into clear and colored portions.

Paper, particularly newspapers and office paper, is recycled to the extent of about 25%. Newsprint does not make large use of reclaimed paper because fibers are degraded in structure and length during processing giving rise to problems in high speed printing [12]. The use of newsprint in cardboard and building applications goes back many years. The degradation of properties and use in downgraded products are factors in polymer use also. The separation problems with paper are simpler than those with polymers since ink and adhesives are not very much like paper pulp. On the other hand, polyethylene and polystyrene have much more in common. Separation of fillers (kaolin and other inorganic materials) from glossy papers is unattractive economically and generally avoided.

Curbside Separations by Consumers.   In the last few years, manufacturers of containers (and some film products) have introduced an identification code which enables consumers to segregate products by polymer type [13]. Thus, almost every bottle and many bags and cups now bear a recycling triangle with number and letter code (Table I).

Table I   **Plastics Identification Code [13]**

Number & Letters (polymer type)	Total	Containers	Film	Typical Uses
	(Sales in billions of pounds [14])			
1 PETE (polyethylene terephthalate)	2.3	1.2	--	Beverages
2 HDPE (high-density polyethylene)	8.5	2.3	0.6	Milk, HIC*
3 V (poly(vinyl chloride))	9.3	0.4	0.3	
4 LDPE (low-density polyethylene)	11.9	0.3	3.7	Bags
5 PP (polypropylene)	8.1	0.3	0.6	
6 PS (polystyrene)	5.1	1.2	0.2	Foam
7 Other	16.3	1.5	0.2	
Totals	61.5	7.2	5.4	

It can be seen that most containers and almost all film materials are potentially included in the 6 categories. There are several cautionary points to be observed. Except for poly(ethylene terephthalate), PET, containers are not the predominant use for any one polymer. For poly(vinyl chloride), PVC, consumer-indentifiable containers are less than 10% of the total. The identification on products is not always very prominently displayed. It can take some effort to ascertain that a given grocery bag is either No. 2 (HDPE), high-density polyethylene, or No. 4 (LDPE), low-density polyethylene. And there is a problem with composite products. Even the symbol on the base of a 2-liter soda bottle is somewhat ambiguous. The symbol is No. 1 (PETE). While it is true that the clear bottle is, indeed, PET, the base (placed over the round bottom of the bottle) is itself No. 2 (HDPE). Also, the bottle usually will contain small amounts of paper (label) and poly(vinyl acetate) (adhesive) in addition to a metal cap with a copolymer liner.

There are other limitations on the system. PET from bottles is of a higher molecular weight than that used for fibers, so PET mixtures will not always be the same on recycling. Whether or not the identification code will be applicable for consumer sorting of more long-lived products like garden hose (PVC) and panty hose (nylon) remains to be seen. Multicomponent products that do not come apart easily (laminated food trays, some food product bottles) end up in the "other" category, No. 7.

On the whole, consumer sorting is a real step forward and greatly simplifies fuller treatment of any type. The environmentally-aware consumer is able to take an active part in a process of great symbolic and some real significance. Because they are such bulky items, removal of aluminum, glass, and plastic containers from the municipal waste stream has a very perceptible impact on any disposal system. The recycling of PET bottles, incidentally, reached over 30% in 1990 [15]. Bottle deposit laws are especially effective in gathering the bottles into central depots.

Post-Consumer Pre-Separation Steps

As an example, we can use the PET beverage bottle since some of the same principles would be applicable to other multi-component waste streams. In fact, the PET bottle has been undergoing an evolution recently. The bottle used here as an example has a metal cap and a polyethylene base. Newer designs use a fluted bottom on an all-PET bottle dispensing with the non-PET base. Moreover, the caps on some bottles now are polyolefinic rather than aluminum. Be that as it may, we shall use the more prevalent (and complex) bottle as our example. We assume here that curbside separation or collection at central locations (like beverage vendors) has already taken place. Recovering PET from bottles involves several basic steps. First, the bottles may be pre-sorted according to color and/or parts, such as base cup, bottle, and cap. This is optional. The bottle must be ground to particle sizes small enough to be processed. After the initial sorting and grinding, each component is sequentially recovered until only PET remains.

Pre-Sorting. There are two reasons for pre-sorting bottles. First, one may separate bottles by color before grinding. The other option is to separate bottles manually or mechanically into major components. Both approaches makes later separations much easier to accomplish. However, this is done at the expense of a slower processing rate and/or a higher labor cost.

One company has reported success in separating green and clear PET without recourse to pre-sorting [16]. At present, most recyclers separate green and clear bottles manually before grinding, if at all.

Manual separation of components can be accomplished by removing the aluminum tops and HDPE bottoms from the PET bottle section, leaving three piles of essentially pure, or at least easily processed, material. There is a machine which separates the HDPE base cups from the rest of the bottle. Unfortunately, this only works on unmutilated bottles [17]. As most bottles are purchased in a crushed and baled state, this limits its use. Dependence on unmutilated bottle condition severely limits available feeds. Increased transportation costs alone would appear to make this an undesirable route to follow. In general, there appear to be technologies available which are capable of handling unsorted bottle components to varying degrees [18,19].

Size Reduction. After any pre-sorting, the scrap must be reduced to a manageable processing size. Shredders and granulators are employed to do this. The standard chip size is 1/4 inch. Chips larger than this size have been determined to cause bridging and plugging within the process. The price for the smaller particle size is a decreasing granulator capacity. If bottles are purchased from a supplier as chip, as sometimes happens [17], only the granulation step is needed. These steps are industry standard.

The granulation of bottles can be responsible for loss of physical strength. The application of mechanical shear in the shredders and granulators breaks the polymer chains into smaller fragments. The decrease in chain size, and thus the molecular weight, can cause a corresponding degradation of the mechanical properties of the polymer. Mechanical properties are responsible for a substantial part of value of most plastics used in films and containers. As long as mechanical reduction of particle size is employed, this is a consideration.

Paper and Metal Separations

From Table II, it is apparent that gravimetric and densiometric methods are the possible approaches to the problem of component separation.

Table II   **Typical Beverage Bottle Components**

Component	Mass		Density
	grams	%	grams/cm^3
Aluminum	1.0	1.1	2.7
HDPE	22	24.2	0.96
Paper	3.9	4.3	1.0
PET	63	69.2	1.36
Poly(vinyl acetate) & copolymers	1.1	1.2	1.2

<u>Gravimetric</u> separation occurs due to a difference in <u>mass</u> between particles. <u>Densiometric</u> separation is more selective as it classifies materials by <u>density</u>, a property specific to each type of component. Size has no effect on densiometric separation. A number of methods exist to separate materials in these ways.

Paper Removal. PET, HDPE, and aluminum cover a particularly wide distribution of densities. Paper, poly(vinyl acetate), and HDPE have densities which are quite similar, though. However, the paper chips are the lightest of all components, even if they are not the most dense. This is because the paper is the thinnest component present; the others are all of rather thick construction.

Being the lightest, the paper is generally removed first. Two approaches are currently practiced. One uses a fluidized bed, and the other uses a cyclone. Both of these approaches are gravimetric. Thus, it should be kept in mind that product loss is possible if polymer particles are of the same mass as the paper. Grinding to too fine a size must be avoided if a clean separation is to be achieved.

In the fluidized bed, air is fed into the bottom and the feed enters near the middle of the bed. When the lifting force of the air overcomes the force of gravity, the particles rise. The lightest particles are pushed out the top and the heaviest sink to the bottom where they are removed. The particles exist in a gradient throughout the column. Several cuts can be taken from the bed.

A cyclone is generally conical in shape, with the feed entering tangentially at the top, the widest part. The outlets are at the bottom and through a pipe whose mouth is lower than the feed entry point. Having the feed enter off-center causes the linear velocity of the stream to be translated into angular velocity. Acceleration rates of several hundred gravities may be easily achieved this way. A vortex is formed, with the lightest material in the center and the heaviest toward the outside. Due to the nature of the vortex formed, material inside the vortex moves upward and out the top of the device. Material outside the vortex moves downwards. In this way separation is simply and effectively accomplished [20]. The cyclone offers simplicity to the process and thus lower capital costs and easier operation.

Fluidized beds offer more flexibility than cyclones in that more than one cut of material may be taken. This is valuable if partial recycling of the product stream is necessary. On the other hand, a second cyclone may be used if the first is not effective enough. Indeed, the use of a sequence of cyclones is a standard approach to many separation operations. Being more complex, fluidized beds are more expensive. There is a trade-off between price and efficiently.

Metal Removal. Two cases must be considered in metal removal: one may assume that aluminum is the only metal present or one may assume that some ferrous contaminants are present as well. If ferrous components are assumed to be present, magnetic separation must be employed [19]. This is generally performed in a drum separator. This would be the case only if very impure feedstocks were being used.

To separate aluminum from plastics, gravimetric, densiometric or electrostatic means can be employed. Gravimetric methods have been less favored, mainly because separation depends upon particle size as well as density. It is quite likely that some metal particles will be lighter than the plastic if only because of their small size. On the other hand, it will always be denser. One firm which has used this approach uses an 'eddy current' separator combined with a fluidized bed to effect removal [18].

Electrostatic separation takes advantage of the difference in the conductivities of materials to sort them. Conducting materials will lose a charge more quickly than insulators. If charged pieces of both materials pass over a grounded drum and then fall past a rotating charged object, the weaker conductor will be more strongly attracted because it will retain more of its original charge. Thus, separation is achieved by the deflection of the insulating fraction away from the conducting fraction. The typical electrostatic separator is based on ion bombardment. An electric discharge forms a corona in which air is ionized. These ions contact the surface of the particles, transferring their charge [21].

Non-Solvent Polymer Separations

High density polyethylene has a specific gravity of about 0.965. The density of the base cup is nearly this although its density can be increased by the amount of pigment (coloring agent) added. It is generally the lightest material remaining when it is removed. The next densest component is PET (1.34-1.39 g/cc) . Polypropylene is an alternative to HDPE in bottle bottoms. However, polypropylene is even less dense than HDPE (s.g. = 0.90 vs 0.96), so all the approaches mentioned will work for both materials.

Direct Flotation. As the HDPE is less dense than water, separation may be easily achieved via direct flotation--a very simple approach if the chips are homogeneous. Direct flotation separates components by density. A mixture is sent to a tank or column which is filled with a liquid whose density is greater than the least dense component. Being less dense than the liquid, the lighter component floats to the top, where it is skimmed off. Since this is a densiometric approach, high purity may be achieved, especially since the difference between the densities of the components is so large. The density of the base cup must be less than that of water, though. Therefore, the amount of pigment added to it must be regulated. To date, however, no problems have been cited.

Froth Flotation. A variation of direct flotation is froth flotation. In froth flotation, a stirred tank is generally employed, with air sparged in from the bottom. The feed also enters from the bottom. Originally a tool of the mining industry, it operates on the principle of the affinity of different materials for air bubbles. Materials with lower wettabilities will cling to air bubbles and rise to the surface. Frothing agents provide a stable froth in which floated materials remain suspended [22]. In the case of PET recycling, though, the wettabilities of the HDPE and PET would be too close for effective separation. Indeed, surfactants are usually added to the direct flotation column to ensure that froth flotation does not occur. Disposal of these chemicals will be discussed later. At present, direct flotation is the preferred approach to HDPE removal.

Water Removal. The presence of water will not only impair the electrostatic separation of PET and aluminum but also severely interfere with any extrusion operations downstream. The maximum water level for extrusion is about 1 ppm [16]. Water levels of less than 0.5 percent must exist for effective electrostatic separation. As washing, rinsing, and flotation are key steps in the PET recovery process, some mechanism for water removal must be present.
 One process uses "spindrying" followed by a dessicant bed to achieve drying [17]. Spindrying presumably refers to centrifugal drying, as is employed by a household washer. The spindrying could also be combined sequentially with heat drying methods. A countercurrent direct heat rotary dryer is a possibility. In this device, feed enters the top of a rotating drum which is tilted at an angle. Hot gases pass up the drum, drying the feed. This is the most thermally efficient of the rotary dryers [23]. Centrifugal drying by itself is more efficient in terms of energy but cannot remove as much water as hot air dryers.

Waste Disposal. All processes must include some form of washing step to remove impurities from the bottles. If the adhesive, poly(vinylacetate), PVA, is to be separated through dissolution or hydrolysis to poly(vinyl alcohol), the waste stream will contain PVA and/or dissolved polyvinyl alcohol in addition to the other contaminants which must be removed. Paper fibers are another major source of impurities. For a 10MM lb/yr plant about 75,000 lb/yr PVA will be removed via the effluent. Environmental regulations probably require that this problem be addressed, especially since polyvinyl alcohol is an excellent foaming agent.

Solvent-Based Separations

We have considered the common non-solvent-based methods of separating post-consumer plastics. However, except for depolymerization, only selective dissolution is capable of purifying bonded, blended, and filled plastics effectively. Dissolution of the polymer releases the impurities which are then removed by filtration, adsorption, or flotation/sedimentation. This yields polymers of high purity for reuse in original applications. The major drawback of a solvent system is the increased expense due to the complexity of equipment and higher energy requirements.

Research on solvent based polymer separation processes is by no means in its infancy. One of the first studies on mixed plastics was conducted by Sperber and Rosen [24,25] in the mid 1970's. These investigators used a blend of xylene and cyclohexanone to separate a mixture of polystyrene (PS), poly(vinyl chloride) (PVC), high density polyethylene (HDPE), low density polyethylene (LDPE), and polypropylene (PP) into three separate phases. In addition, many United States and foreign patents dating from the 1970's were granted for the solvent recovery of thermoplastic polymers [26-30]. The interest in solvent processes waned in the late 1970's as the oil crisis eased, but the growing need to develop solutions to the solid waste problem has renewed the research effort [31-33].

Representative Processes. Early workers on solvent processes for mixed plastics [24,25] could not foresee the penetration of poly(ethylene terephthalate) (PET) into the packaging industry and, so, their processes did not address the separation of PET from mixtures. The RPI process [33], a more recent research project on mixed plastics, does include the separation of PET. This process utilizes a single solvent (either tetrahydrofuran or xylene) to separate the polymers in batch mode based on the temperature dependent dissolution rates of each resin. The polymer solution is then exposed to elevated temperatures and pressures before flash devolatilizing the solution. In addition to the high pressures required in the feed to the flash stage, elevated pressures are required in the dissolution stage of the RPI process because the normal boiling point of the solvent is below the temperature required to achieve PET dissolution. The use of elevated temperatures increases the risk of thermally degrading the polymers while the elevated pressures translate into higher energy and equipment costs. An additional limitation of the RPI process lies in the use of a single solvent capable of dissolving all of the polymers. This reduces the purity of the final polymer streams due to the partial dissolution of polymers which were intended to be dissolved at a higher temperature and the carryover of undissolved polymers to a higher temperature dissolving fraction. These limitations can be overcome by using multiple solvents with each solvent being compatible with only a limited number of polymers at reasonable temperatures and by incorporating a less energy intensive recovery stage. In addition, lower cost processes could be employed to make initial separations.

Combined Technology Process. Each separation process has its merits and limitations. It would seem logical that a process combining various separation technologies could take advantage of the inherent strengths of an individual process, while attenuating the limitations. For example, a combined technology process could utilize a sink-float system to produce segregated polymer streams which could then be further purified in solvent-processing trains. The strength of the sink-float system is cost-effectiveness, but it is limited by the relatively low purity of the final products, as is true of all compositionally blind separators. Selective dissolution processes, on the other hand, can yield polymers of high purity but at an increased cost.

Example - 2-liter Bottle Process. This concept of integrated technology can be applied to a waste stream of 2-liter bottles composed of PET, HDPE base cup, paper or

PP label, PP or aluminum cap, cap liner, adhesives, and associated fillers, catalysts, pigments, and impurities. A simplified process diagram of such a process is shown in Figure 1. The bottles are shredded, air classified to remove paper, washed with a detergent/water mixture, then sent to a sink-float tank containing water. In this tank, the less dense polyolefins (PP and HDPE) are separated from the denser aluminum and PET yielding streams which will be less susceptible to cross-contamination due to unintended polymer dissolution in the solvent stages. This contamination is not eliminated, however, since the sink-float system cannot be 100% efficient. After removal from the water tank,, the streams are dried and fed to separate solvent processing trains.

A flow diagram of a typical solvent process is displayed in Figure 2. Dried chips from the sink-float system are fed to Stage I where they are washed with the process solvent at a temperature sufficient to remove soluble impurities, but insufficient to dissolve the target polymer. For example, in the PET train, this washing will remove any adhesive, PS, or PVC which may be present from use in 2-liter bottles or from sortation errors. The solvent-washed chips then enter Stage II where the target polymer is dissolved by the process solvent at higher temperatures. Once dissolution is complete, the solution can be purified (Stage III). In this stage, undissolved materials are removed utilizing sedimentation/flotation and filtration and dissolved materials, such as dyes, catalysts, and inks, are removed by adsorbents. The next stages of the process consist of polymer recovery (Stage IV) by temperature quenching or non-solvent addition and rinsing (Stage V) to remove dissolved impurities from the polymer precipitate. Finally, the solvent is removed (Stage VI) followed by polymer extrusion and pelletization (Stage VII).

The selection of solvents for the treatment trains is based on the thermodynamics and dissolution/precipitation kinetics of polymer-solvent pairs as well as the toxicity, cost, and solvent recovery feasibility. By selecting solvents which are optimal for each polymer stream, cross-contamination is minimized and expenses reduced. In this way, polymer/solvent compatibility is achieved by solvent selection rather than elevated pressures and temperatures.

The initial solvent of choice for the polyolefin treatment train is xylene. Much research has been published on the dissolution and precipitation behavior of polyethylene and polypropylene in xylene [34-39]. In addition, researchers at RPI have shown that xylene can be used to separate PP from HDPE in a solvent process [33]. Although it is apparent that xylene is acceptable for this treatment, additional research aimed at separating these specific polymers may yield a more suitable solvent.

PET Processing Train.   The major focus of the present research [40] on this combined technology example has been in the selection and evaluation of solvents for the PET treatment train. Based on the selection criteria of cost, toxicity, solvent recovery, favorable PET solution behavior, and incompatibility with polyolefins, the industrial solvent n-methyl-2-pyrrolidinone (NMP) was chosen. NMP is used in many processes [41], is of low toxicity [42,43], and is easily recovered [41]. In addition, NMP is capable of completely dissolving PET while only dissolving minute amounts of the polyolefins.

In order to determine the appropriate operating conditions for the PET processing train, the dissolution and crystallization behavior of PET in NMP had to be studied. The dissolution rate of PET in NMP as a function of solvent temperature was established by measuring the mass loss of 2-liter bottle sections (two bottles) when exposed to NMP at various temperatures and times. The results of this study indicate that at 130°C the PET sections only swelled in the solvent, and did not dissolve appreciably even after over 4 hours in the solvent. At 140°C, however, the sections dissolved completely in less than 10 minutes. Figure 3 illustrates the linear increase in dissolution rate as temperature increases above   140°C   (presented as % dissolved per

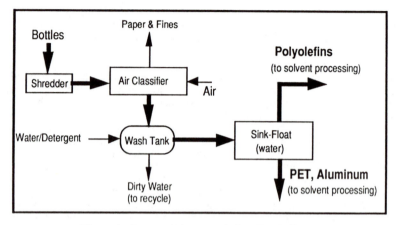

Figure 1. Process Diagram: 2-liter Bottle Waste

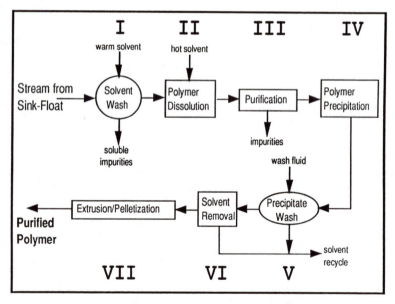

Figure 2. Flow Diagram: Solvent Process

minute). From Figure 3 it is apparent that PET dissolution is achievable without elevated pressures since the normal boiling point of NMP is 202°C. These data indicate that the temperature of Stage I (solvent rinse) of the solvent processing train can be as high as 130°C, thereby dissolving any remaining adhesives, PS, and PVC while leaving the PET only slightly swollen. The NMP in Stage II (polymer dissolution) should be at a temperature above 160°C in order to take advantage of the favorable kinetics. The dissolution experiments showed that at 170°C, only ≈4 minutes are necessary for complete dissolution of the 2-liter bottle sections. A PET concentration of 10 weight% (wt%) is advisable in Stage II to avoid processing complications caused by increased solution viscosity. However, PET concentrations as high as 30 wt% PET are achievable.

After purification of the PET/NMP solution, the polymer is recovered either by temperature quenching the solution to bring about the crystallization of PET or by adding the solution to a non-solvent causing immediate precipitation of the polymer. The quench crystallization of PET from an NMP solution is a nucleation controlled process and is, therefore, extremely temperature sensitive. In addition, the rate is dependent upon the polymer concentration. In order to study this phenomenon, samples were prepared from bottle grade PET (Goodyear Cleartuf 7207) and anhydrous NMP (Aldrich) in flame sealed, evacuated, glass cells. The time required to initiate crystallization as a function of quench temperature and polymer concentration is shown in Figure 4 and Figure 5. This data was obtained from plots of sample transmittance as a function of time. From Figures 4 and 5 it is apparent that crystallization becomes rather rapid as the quench temperature is decreased and the polymer concentration is increased. The product of a quench crystallization is a suspension of PET which can be rinsed with another liquid, such as acetone, to remove the bulk of the NMP followed by drying. While a slow crystallization may appear unattractive at first, it can be of benefit since impurities might precipitate before PET and the solution could be filtered at the quench temperature before the PET is able to crystallize.

The alternative to temperature quenching is shock precipitation brought on by mixing of the PET solution with a non-solvent. The characteristics of the non-solvent and mixing conditions are of great importance since they affect the structure of the precipitate. For example, addition of a PET solution to xylene or acetone with mixing yields a finely dispersed precipitate which is easier to concentrate by filtration than a quench crystallized sample. On the other hand, if the solution is injected into a water bath (wet spun) with only slight agitation, a strand of gelled precipitate is formed. The relative ease of solvent removal from the finely dispersed precipitate makes this type of shock recovery more attractive than the formation of gelled strands.

Impurities. Like all processes, the PET processing train is limited by the presence of impurities. The main impurities in the train are aluminum, HDPE, and PP. Aluminum removal is extremely efficient in the solvent process because it can be easily settled and/or filtered out of the polymer solution. This alleviates the need for eddy current separators which represent a large fraction of plant costs for other processes [44]. Because the sink-float system is not 100% effective, small amounts of the polyolefins will make their way into the PET processing stages. The advantage of using NMP in this system is that the polyolefins are only slightly soluble at elevated temperatures (probably low molecular weight fractions). The extent of dissolution of HDPE in NMP as a function of time and temperature is shown in Figure 6. The data indicate that a small fraction of the HDPE dissolves. For example, merely 0.2 % of an HDPE sample exposed to NMP at 175°C for 30 minutes will dissolve.

If a 2-liter bottle waste is assumed to be composed of 70 wt% PET and 30 wt% HDPE and this waste is exposed to NMP at 175°C for 30 minutes followed by flotation/sedimentation, filtration, and precipitation, the resulting polymer would be 99.91 wt% PET. If the NMP treatment had been preceded by a sink-float stage capable

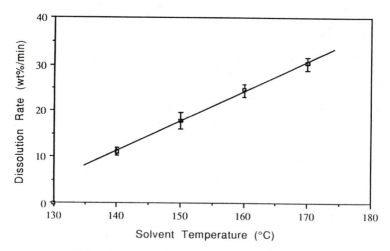

Figure 3. Dissolution Rate of Bottle PET in NMP
as a Function of Temperature

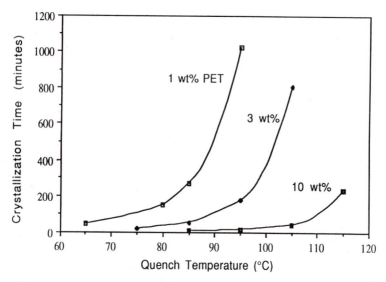

Figure 4. Effect of Quench Temperature and PET Concentration
on Crystallization Time for PET in NMP (Goodyear Cleartuf 7207 PET)

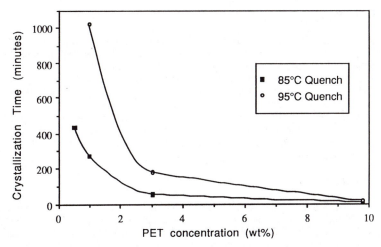

Figure 5. Effect of PET Concentration and Quench Temperature
on Crystallization Time for PET in NMP (Goodyear Cleartuf 7207 PET)

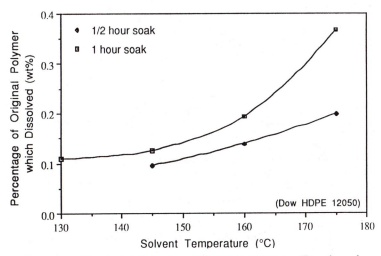

Figure 6. Dissolution Behavior of HDPE in NMP as a Function of
Solvent Temperature and Soak Time

of removing only 90% of the HDPE, the polymer product would be 99.991 wt% PET. Likewise, 99.9991 wt% PET is achieved with a 99% sink-float efficiency followed by NMP processing. While this exercise overlooked all other impurities, it should be evident that a final product with a purity on the order of 99.99 wt% PET is within reason. This level of purity and the sterilizing conditions of the process could make the product a candidate for original use applications.

Conclusions

A combined technology process is an attractive approach to solving the mixed waste problem. The polymer streams from such a process should be appropriate for high grade applications, with only a fraction of the waste relegated to low grade, non-critical applications. An efficient scheme for implementing the 2-liter bottle example process would include curbside collection, container sortation (a mixed stream of PET and HDPE containers is acceptable), and shredding to increase the bulk density of the plastic, thereby reducing the shipping costs. The latter steps would be accomplished at a Material Recovery Facility (MRF). A group of MRF's could then sell the shredded containers to a centralized combined technology facility where the sink-float and solvent stages will be located.

Acknowledgments

The section on Solvent-Based Separations is based in part on a paper presented at the Regional Technical Meeting on Plastic Waste Management sponsored by the Society of Plastics Engineers, Oct. 17-18, 1990. The authors wish to thank Dow and Goodyear for polymer samples. Also, L. M. Vane thanks the Plastics Institute of America for a supplemental fellowship, and the National Science Foundation and DuPont for stipend support.

Literature Cited

1.   *Waste Solutions*, Supplement to *Mod. Plastics*, April, 1990.
2.   Nir, M. M., *Plastics Eng.*, **1990**, *46 (Sept.)*, 29; *(Oct.)*, 21.
3.   Huang, J-H., Shetty, A. S., and Wang, M-S., *Adv. Polym. Tech.*, **1990**, *10*, 23.
4.   Thayer, A. M., *Chem. & Eng. News*, **1990**, *69 (June 25)*, 7.
5.   Brunner, C. R., *Chem. Eng.*, **1987**, *94 (Oct. 12)*, 96.
6.   Kaminsky, W., *J. Anal. & Appl. Pyrolysis*, **1985**, *8*, 439.
7.   Rodriguez, F., *Principles of Polymer Systems, 3d ed.*,Hemisphere, New York, NY, 1989, p. 331.
8.   Ibid., p. 338.
9.   *Metal Statistics 1989*, Fairchild Publications, New York, NY, 1989, pp. 26-28.
10.   Carlin, J. F., in *Minerals Yearbook*, U. S. Dept. of Interior, Bureau of Mines, Washington, DC, 1990, Vol. 1, p. 962.
11.   *Statistical Abstract of the United States*, U. S. Dept. Commerce, Washington, DC, 1990, p. 206.
12.   Leaversuch, R., *Mod. Plastics*, **1988**, *65 (June)*, 65.
13.   *Plastic Container Code System*, The Plastic Bottle Information Bureau, Washington, DC.
14.   *Mod. Plastics*, **1991**, *68 (Jan.)*
15.   *Mod. Plastics*, **1991**, *68 (April)*, 38.
16.   *Chem. & Eng. News*, **1985**, *63 (Oct. 21)*, 26.
17.   Smoluk, G. R., *Mod. Plastics*, **1988**, *65 (Feb.)*, 87.
18.   *Mod. Plastics*, **1980**, *57 (April)*, 82.
19.   *Chem. Eng.*, **1984**, *91 (June 25)*, 22.

20.   *Perry's Chemical Engineers Handbook*, 6th ed., Perry, R. H., and D. W. Green, Eds. McGraw-Hill, New York, 1986, p. 21-32.
21.   *Ibid.*, pp. 21-42. through 21-44.
22.   "Froth Flotation," in *Kirk-Othmer Encyclopedia of Chemical Technology, 3rd ed.*, Wiley, New York, 1982, Vol. 18.
23.   *Environmental Impact of Nitrile Barrier Containers, LOPAC: A Case Study*, Monsanto Co., St. Louis, MO, 1973, p. 24.
24.   Sperber, R. J., and Rosen, S. L., *Polym. Eng. Sci.*, **1976**, *16*, 246.
25.   Sperber, R. J., and Rosen, S. L., *SPE ANTEC Tech. Papers*, **1975**, *21*, 521.
26.   Meyer, M. F.,Jr. *et al.* (to Eastman Kodak), US Pat. 3,701,741 (Oct 31, 1972).
27.   Kajimoto, H., *et al.* (to Mitsubishi Heavy Industries) Japan. Kokai Pat. 76 16,378 (Feb 9,1976).
28.   Sidebotham, N. C., *et al.* (to Monsanto), US Pat. 4,003,881 (Jan 18,1977).
29.   Nishimoto, Y., *et al.* (to Mitsubishi Heavy Industries), Japan. Kokai Pat. 76 20,976 (Feb 19,1976).
30.   Sidebotham, N. C., *et al.* (to Monsanto), US Pat. 4,137,393 (Jan 30, 1979).
31.   Kampouris, E. M., Papaspyrides, C. D., and Lekakou, C. N., *Polym. Eng. Sci.*, **1988**, *28*, 534.
32.   Drain, K. F., Murphy, W. R., and Otterburn, M. S., *Conservation and Recycling*,**1983**, *6*, 107.
33.   Lynch, J. C., and Nauman, E. B., presented at SPE RETEC: New Developments in Plastic Recycling, Oct. 30-31, 1989.
34.   Prasad, A., and Mandelkern, L., *Macromolecules*, **1989**, *22*, 914.
35.   Mandelkern, L., *Polymer Preprints*, **1986**, *27*, 206.
36.   Devoy, C., Mandelkern, L., and Bourland, L., *J. Polym. Sci.* A-2,**1970**, *8*, 869.
37.   Mandelkern, L., *Polymer*, **1964**, *5*, 637.
38.   Jackson, J. F., and Mandelkern, L., *Macromolecules*, **1968**, *1*, 546.
39.   Drain, K. F., Murphy, W. R., and Otterburn, M. S., *Conservation and Recycling*,**1981**, *4*, 201.
40.   Vane, L. M., Ph.D. Thesis, Cornell University, Ithaca, NY, 1991.
41.   *M-PYROL, N-Methyl-2-Pyrrolidone HANDBOOK*, GAF Corp., New York, NY, 1972.
42.   Lee, K. P., *et al.*, *Fundam. Appl. Toxicol.*, **1987**, *9*, 222.
43.   Rapid Guide to Hazardous Chemicals in the Workplace, Sax, N. I., and Lewis, R. J., Sr., Eds.Van Nostrand Reinhold Co., New York, NY, 1986.
44.   Thayer, A. M., *Chem. & Eng. News*, Jan. 30, 1989, p. 7.

RECEIVED April 1, 1992

WASTE TREATMENT AND AVOIDANCE

Chapter 9

Removal of Hydrophobic Organic Compounds from the Aqueous Phase

Continuous Non-Foaming Adsorptive Bubble Separation Processes

K. T. Valsaraj, X. Y. Lu, and L. J. Thibodeaux

Department of Chemical Engineering, Louisiana State University, Baton Rouge, LA 70803

The steady state separation of three hydrophobic organic compounds from the aqueous phase by two non-foaming adsorptive bubble techniques was studied. The two processes were bubble fractionation and solvent sublation carried out in continuous countercurrent modes of operation. The hydrophobic compounds considered for removal were naphthalene, pentachlorophenol and 2,4,6- trichlorophenol. Mineral oil was used as the organic solvent in sublation. A small scale 5-cm diameter glass column and a large scale 15-cm diameter plexiglass column were used in the experiments. The effects of several process variables such as air flow rates, influent feed rates, organic solvent volumes (in sublation) and effluent flow rates (in bubble fractionation) were studied. Increasing column diameter had a dramatic effect on the degree of re-dispersion of solute at the top of the aqueous section in bubble fractionation whereas there was little effect in the case of solvent sublation. The steady state fractional removal in both bubble fractionation and solvent sublation were primarily dependent on the average air bubble size in the aqueous phase.

A number of processes that utilise the surface active nature of hydrophobic compounds in separating them from the aqueous phase are described in the literature. Of these the so-called *adsorptive bubble separations* form a class of techniques that utilise air bubbles to concentrate and separate surface active compounds (*1*). There are two main classifications of these, viz., foaming and non-foaming separations (*2*). The focus of attention has mostly been on foam separations. However, in those cases that involve low concentrations of

0097–6156/92/0509–0116$06.00/0

surfactants where practically no stable foam is possible, non-foaming separations are useful. However, their potential has not been fully realised as yet. Only two processes fall under this category- *bubble fractionation* and *solvent sublation*. In spite of the fact that these processes were developed in the 1960's and 1970's, very little work has been done on these processes as far as large scale applications are concerned. Most of the work are on lab scale semi-batch and continuous columns (*1,3*). Although lab scale columns are useful in understanding the mechanisms of these processes, they are of little value in ascertaining the utility of these processes on large scale apparatus. Some recent studies in bubble fractionation have shown that due to increased axial dispersion in large columns, the process efficiency decreases rapidly with scale up (*12*).

Both solvent sublation and bubble fractionation have similar mechanisms of solute transport by air bubbles. In bubble fractionation, the surface active material transported by the air bubble is deposited at the top of the aqueous phase as the bubbles burst thus enriching the upper section at the expense of the lower section (*5*). If a continuous overflow of the liquid in the upper section is considered, as in a continuous process (Figure 1) with influent fed at some point along the aqueous section and effluent drawn at the bottom, one can reduce the re-dispersion of material from the enriched upper section and obtain a good separation efficiency. The separation action is primarily opposed by the axial dispersion in the aqueous phase. In continuous bubble fractionation, we have an enriching section above the feed point and a stripping section below the feed point (*4*). In solvent sublation the same mechanism is operative. However, usually the entire aqueous phase is used as a stripping section with no liquid overflow at the top of the aqueous section. Instead an immiscible organic solvent is floated on top of the aqueous section to collect and extract the material brought to the surface by the air bubbles (Figure 1). The organic solvent may or may not be in continuous flow. Most experiments to date have used a stagnant solvent layer. The sublation mechanism involves (i) a unidirectional transport of the hydrophobic organic material by the air bubbles across the aqueous-organic phase interface and, (ii) an extraction of this material into the organic phase by molecular diffusion (*6*). The axial concentration gradient in the aqueous section is counteracted by axial dispersion caused by local circulations set up by the air bubbles and gross circulations of the liquid in the column. The requirements for both bubble fractionation and solvent sublation are small air bubbles (to maximise the air-water interfacial area) and tall, narrow bubble columns (to reduce axial dispersion in the aqueous phase). Theoretical work on mechanisms and mathematical models for steady state continuous bubble fractionation were proposed by Bruin *et al.*(*4*). Valsaraj and Thibodeaux (*6,7,10*) have investigated the mechanisms of solvent sublation in the semi-batch mode and have proposed mathematical models for both semi-batch and continuous solvent sublation. The advantages of solvent sublation over solvent extraction and conventional air stripping were also investigated by Valsaraj and co-workers (*6,7,10*).

In the present study the removal of three hydrophobic compounds of environmental significance was explored using two bubble columns of different diameters utilising solvent sublation and bubble fractionation in the countercurrent mode. The fractional removals in the stripping sections of the aqueous phase for the two processes were obtained. The effects of air flow rate, aqueous feed rate, and column diameter on the fractional removals at steady state in both processes were studied. The effects of organic solvent volume in solvent sublation were also studied along with the effect of the top overflow rate in bubble fractionation. This study was undertaken to elucidate the feasibility of these two separation operations in the continuous countercurrent mode. Most of the available data are in the semi-batch mode and do not provide adequate information as to the possible problems in the scale up of these operations. One of the aims of this study was to identify any limitations on the scale-up of non-foaming separations.

EXPERIMENTAL

The bubble columns used were a 5-cm diameter, 100-cm tall glass column and a 15-cm diameter, 150-cm tall plexiglass column. The column configuration and ancillaries are described elsewhere (6). Air bubbles were generated by passing house air through a medium porosity Pyrex brand fritted glass sparger in the 5-cm diameter glass column and a fine porosity fritted glass sparger in the 15-cm diameter plexiglass column. The bubble columns were mounted perfectly vertical to minimize the wall effect on axial dispersion (8,9). Aqueous feed was pumped into the columns using Model 403 Peristaltic pumps (Scientific Industries, Bohemia, NY) at flow rates from 6.3 to 25 mL.min^{-1}.

In solvent sublation experiments the aqueous feed was introduced using a liquid distributor in the aqueous phase. The organic solvent on top of the aqueous phase was stagnant. In bubble fractionation experiments the top and the bottom effluent rates were monitored and adjusted as required. The aqueous feed in bubble fractionation experiments was introduced using the liquid distributor. The overall operation in both cases was thus one in which the aqueous phase was in continuous countercurrent contact with air. The air was in a once-pass-through mode. The air flow rate was measured using a soap bubble flowmeter and a rotameter.

The organic solvent used in the solvent sublation experiments was a light mineral oil supplied by Fisher Scientific Co. It is immiscible with water, non-toxic, and has high affinity for the hydrophobic compounds considered for removal.

Three hydrophobic compounds were chosen for these experiments, viz., pentachlorophenol (PCP), 2,4,6-trichlorophenol (TCP), and naphthalene (NAPH). PCP and TCP were supplied by Aldrich Chemicals and NAPH was supplied by Fisher Scientific. The choice of these compounds was made based on a variety of factors including vapor pressure, aqueous solubility, and hydrophobicity as indicated by their octanol-water partition constants. All of these properties spanned a wide range of values as shown in Table I.

PCP and TCP exist as neutral hydrophobic species at pH values below their pK_a. By conducting all experiments at a pH of 3.0, we focussed our attention on the removal of only neutral PCP and TCP species. Both these species have very low Henry's constants (H_c) and high air-water interfacial

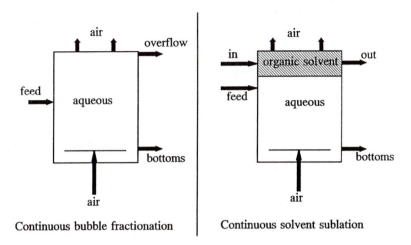

Continuous bubble fractionation | Continuous solvent sublation

Figure 1. Schematic of continuous bubble fractionation and solvent sublation.

Table I. Relevant Properties of Compounds Used[c](7,10)

PROPERTY	PCP[a]	TCP[a]	NAPH
Molecular weight,g/mol	266	197	128
Aqueous solubility,mg/L	14	800	34
Vapor pressure,mm Hg	1.95×10^{-5}	0.015	0.064
Henry's constant,--	2×10^{-5}	2×10^{-4}	1.7×10^{2}
Adsorption constant,cm	1.8×10^{-4}	8×10^{-5}	1.0×10^{4}
log octanol-water partition constant,--	5.01	3.38	4.17
pK_a	$4.7^{(b)}$	$6.0^{(b)}$	-----
Mineral oil-water partition constant,--	490	19	-----

Note: [a]For PCP and TCP, the properties given are for neutral species only.
[b]pK_a = -log K_a, where K_a is the ionisation constant of the acid species.
[c]All properties are at 25°C.

adsorption constants (K_d) while NAPH has a large H_c value as well as a high K_d value. The Henry's constant is defined as

$$H_c = \frac{vapor\ concentration,\ mol.cm^{-3}}{aqueous\ concentration,\ mol.cm^{-3}} \tag{1}$$

and the air-water interfacial adsorption constant is defined as

$$K_d = \frac{air-water\ interface\ concentration,\ mol.cm^{-2}}{aqueous\ concentration,\ mol.cm^{-3}} \tag{2}$$

Since the separation factor depends on the value of ($H_c + 6\ K_d/d$), it is clear that the removal of PCP and TCP occurs predominantly by adsorption on the air bubble surface while NAPH removal occurs by transfer inside the air bubbles.

Aqueous solutions of the above compounds were prepared by overnight stirring and subsequent filtration to remove any suspended material. The concentrations in the aqueous feed and bottom effluent were measured at steady state which was reached in about 4 hours in most experiments (solvent sublation and bubble fractionation). The fraction removed at steady state is reported as R where R is given by 1 - (C_e/C_i) where C_e is the bottom effluent concentration and C_i is the feed concentration (see Figure 1 for details). After acidification of the samples collected from the bottom effluent, concentrations of TCP and PCP were monitored by measuring their UV absorbance on a UV-visible spectrophotometer (HP 8452A with diode array detector) at their primary wavelength maxima of 211 nm and 214 nm, respectively. Concentrations were obtained by comparison with calibration standards. NAPH concentration was determined using a gas chromatograph. The gas chromatograph used was a Hewlett Packard HP 5890A equipped with a flame ionization detector. The aqueous samples were injected directly into the GC using Hamilton syringes. The column used was 6 ft. long, had 1/8 inch internal diameter, was made of stainless steel, and was packed with 3% SP-2250 on 80/120 Supelcoport (Supelco Inc.). The detector and injector temperatures were 220 and 220°C respectively. The oven temperature was set for an initial value of 100°C, a ramp of 15°C per minute and a final temperature of 130°C.

RESULTS AND DISCUSSION

The analysis of the data from a continuous steady state operation was described by Bruin et al. (4) for bubble fractionation and by Valsaraj and Thibodeaux (7) for solvent sublation.

According to Bruin et al. (4) who proposed a mathematical model for continuous bubble fractionation, a convenient measure of the fractional removal is given by the ratio

$$R_{BF} = \frac{WC_T - WC_i}{LC_i - WC_i} \qquad (3)$$

The above equation gives the ratio of the difference between the actual flow of solute in the overflow ($= W \, C_T$) and the overflow if no separation was achieved ($= W \, C_i$) to the maximum difference if complete separation was achieved in the overflow ($= L \, C_i - W \, C_i$). Since $L = E + W$ and $LC_i = WC_T + EC_e$, one obtains

$$R_{BF} = 1 - \frac{C_e}{C_i} \qquad (4)$$

It should be noted that the above equation gives the efficiency of the stripping section in the aqueous column (Figure 1).

In the case of continuous solvent sublation the fractional removal in the aqueous phase at steady state is given by the ratio (7)

$$R_{SS} = 1 - \frac{C_e}{C_i} \qquad (5)$$

The above equation for solvent sublation applies since the entire aqueous section is used as a stripping section (Figure 1).

Using Equations (4) and (5) the fractional removals in the *stripping section* of the aqueous phase in a bubble fractionation column and a solvent sublation column were obtained.

Figure 2 gives the steady state removal for the three compounds NAPH, PCP and TCP by solvent sublation on the small scale 5-cm diameter column. The total aqueous height was 90 cm and the mineral oil height was 1.2 cm for NAPH sublation and 0.3 cm for both PCP and TCP sublation. The aqueous feed rate was fixed at 6.3 mL.min^{-1} (loading rate of 0.38 cm^3.cm^{-2}.min^{-1}) and the mineral oil layer was stagnant during the course of the experiment. The aqueous influent was fed at a height of 80 cm above the air sparger.

Figure 3 gives the results for the steady state removal of the three compounds using bubble fractionation on the small scale 5-cm diameter bubble column. The aqueous influent feed rate was 6 mL.min^{-1}. The total aqueous height was 90 cm and the influent was fed at a height of 50 cm above the air sparger. The bottom effluent was maintained at 3 mL.min^{-1} and hence the overflow rate was 3 mL.min^{-1}.

Figures 2 and 3 show that for both bubble fractionation and solvent sublation, the fraction removed at any particular G/L is in the order NAPH > PCP > TCP. According to the models for bubble fractionation (4) and solvent sublation (3) the fractiuonal removal from the aqueous phase is primarily dependent on the "separation factor" S, given by $(H_c + 6K_d/d)(G/L)$ where d is the average bubble diameter. If G/L is held constant the degree of removal for any particular compound depends uniquely on the "effective partition constant",

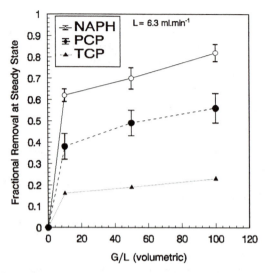

Figure 2. Effect of air/water volumetric flow ratio on the solvent sublation of PCP, TCP and NAPH on the small-scale 5 cm diameter column.

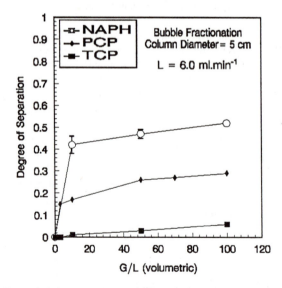

Figure 3. Effect of air/water volumetric flow ratio on the bubble fractionation of PCP, TCP and NAPH on small-scale 5 cm diameter bubble column.

($H_c + 6K_d/d$) between the air bubble and the aqueous phase. If a constant d of 0.1 cm is assumed and on utilizing Table I for the other parameters, we obtain values of ($H_c + 6K_d/d$) of 0.023, 0.011 and 0.005 for NAPH, PCP and TCP respectively. It is thus apparent that the degrees of separation are in the same order as the effective partition constants (or the separation factors) and are in accordance with the theoretical models. Previous work on semi-batch processes for both bubble fractionation and solvent sublation have led to similar conclusions (*4,10*).

A particularly noteworthy conclusion from Figures 2 and 3 is that the fractional removal in the stripping section for solvent sublation is higher than for bubble fractionation. Several reasons can be ascribed to this observation. Firstly, solvent sublation is a combination of transport by bubbles (as in bubble fractionation) and simultaneous solvent extraction. However, in bubble fractionation the only mechanism of solute removal at the top of the aqueous section is by liquid overflow. This overflow will be incomplete in most cases (*12*) and considerable backmixing by axial dispersion into the bulk aqueous phase is likely, thereby decreasing the fractional removal in bubble fractionation. Other investigators have noted that by increasing the top overflow rate and decreasing the bottom effluent flow rate while maintaining a constant influent feed rate, one can obtain increased fractional removals in bubble fractionation (*4,12*). In solvent sublation a high fractional removal is possible from the top of the aqueous section into the organic solvent phase with reduced backmixing, particularly when the compound has a high affinity for the organic solvent (i.e., a large partition constant for the solute) and when the organic solvent volume is large enough so as to have a high capacity for the solute.

The relative magnitudes of H_c and K_d (Table I) indicate that the removal of both PCP and TCP occurs mainly by adsorption on the air-water interface of the air bubbles while that of NAPH occurs mostly as the vapor phase inside the air bubbles. Since the volatile fraction carried inside the air bubbles is lost to the air at the top of the aqueous phase and the surface adsorbed phase is deposited at the top of the aqueous phase, bubble fractionation removal of NAPH involves simultaneous gas desorption (air stripping) and enrichment at the top.   Bubble fractionation is an integral part of most operations conducted in bubble columns.

The effect of the feed point location in bubble fractionation was marginal. For example, increasing the stripping height from 50 cm (as in Figure 3) to 74 cm changed the degree of separation for PCP from 0.17 to 0.22 at a G/L of 10, from 0.26 to 0.27 at a G/L of 50 and from 0.27 to 0.28 at a G/L of 70. We interpret this to mean that a height of 50 cm or larger is enough to attain a large number of equilibrium stages in the aqueous phase and increasing it further will not effect the degree of separation to any large extent. Previous work lend support to this fact (*4*).

The effect of mineral oil volume on the degree of separation in solvent sublation was not significant. For example, at a constant aqueous depth and aqueous influent feed rate, an increase in organic solvent volume from 5 mL (0.3 cm depth) to 15 mL (0.9 cm depth) increased the steady state degree of separation for PCP from 0.38 to 0.42 at a G/L of 10 and from 0.49 to 0.50 at a G/L of 50. Just as in solvent extraction, beyond a certain organic solvent volume, the degree of separation in solvent sublation is also independent of organic solvent volume as has been demonstrated by us earlier (6).

Figures 2 and 3 show that the degree of separation initially increased with increasing G/L ratio at a constant feed rate L, and approached an asymptotic value at large G/L values (i.e., large air flow rates, G). This is understandable, since at a low air flow rate the value of the average (Sauter mean) bubble diameter was small and increased with increasing air flow rate as shown in Table II. The Sauter mean bubble diameter was obtained using a video image analysis of the column in operation at various air flow rates (10). The increased average bubble diameter meant decreased surface area per unit volume of air available for adsorptive separation and hence the degree of removal did not increase with increasing air flow rates. Previous work has shown a similar behavior for semi-batch experiments(7). The approach to an asymptotic degree of separation at high G/L values indicates that the fractional removals in both bubble fractionation and solvent sublation are sensitive to the average bubble size. Small bubble sizes and low gas flow rates are therefore essential prerequisites for good, efficient separation with the type of gas injection method (i.e., a porous gas sparger) that was used in these experiments. Since at high air flow rates, the increase in bubble size is mainly due to coalescence of bubbles near the gas sparger, improvements in the degree of separation can be achieved by using methods such as those suggested by Wilson and Pearson(11). The above limitations point to the need for devising alternate ways of producing and maintaining small air bubbles even at high air flow rates.

Figure 4 gives the fractional removal for steady state solvent sublation of PCP on the large 15 cm diameter column with a total aqueous depth of 150 cm. The effects of both mineral oil volume and influent feed rate are shown. The effect of increased solvent volume seems to be insignificant while increasing influent feed rate decreased the degree of separation considerably. The decrease in fractional removal is far more dramatic at higher G/L values. Increasing feed rate,L from 6.3 to 25 mL.min^{-1} decreased the degree of separation from 0.30 to 0.17 at a constant G/L of 10 while it decreased from 0.58 to 0.19 at a G/L of 50. As discussed earlier for the small scale column, because of increased air bubble diameter with increase in air flow rates the fractional removal reached an asymptotic value at large G/L values. It is clear that the effects of increased countercurrent liquid flow rates on the degree of separation in solvent sublation cannot be offset by increased gas flow rates and hence form a severe limitation of solvent sublation.

The fractional removal at steady state in bubble fractionation of PCP on the 15 cm diameter column is shown in Figure 5. The total feed rate L, was 22 mL.min^{-1} in the first series of experiments. The bottom effluent rate, E was 11 mL.min^{-1} and the overflow rate,W was 11 mL.min^{-1} with a height of the stripping section maintained at 80 cm. The fractional removal increased from 0.11 at a G/L of 10 to only 0.13 at a G/L of 100 . The fractional removal was dependent on the ratio E/L. When E/L decreased, the degree of separation increased as shown in Figure 5. This is in accord with theory which predicts that R should increase with decrease in E/L, since a decrease in E increases the "separation

Table II. Air Flow Rate and Sauter Mean Bubble Diameter(*10*)

Air flow rate (mL/min)	Sauter Mean Bubble Diameter (cm)	
	In pure water	In water satd.w/PCP
60	0.117±0.038	0.091±0.028
180	0.152±0.050	0.108±0.025
300	0.170±0.054	0.131±0.034
420	0.199±0.062	0.146±0.039
600	0.216±0.057	--

Note: Mean and standard deviations are given. Sauter mean bubble diameter is the volume to surface averaged bubble diameter in a bubble swarm.

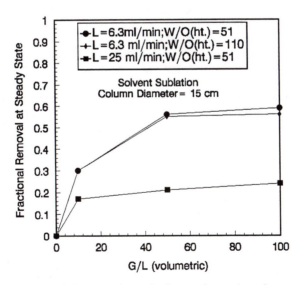

Figure 4. Effect of air/water volumetric flow ratio on the solvent sublation of PCP on the large-scale bubble column.

factor" for the stripping section in a bubble fractionation process(4). The effect of increased liquid loading on fractional removal appeared to be more dramatic at high G/L values just as was observed for the solvent sublation experiments described earlier. When the fractional removals in bubble fractionation process at a feed rate of 20 mL.min^{-1} and bottom effluent rate of 5 mL.min^{-1} in Figure 5 were compared with the fractional removals in solvent sublation experiments at a feed rate of 25 mL.min^{-1} in Figure 4, we observed that the fractional removal in bubble fractionation approached that of solvent sublation as the overflow rate in bubble fractionation was increased. With increased overflow rate in bubble fractionation the re-dispersion of solute from the top aqueous section would be minimised and hence should lead to an increase in the fractional removal.

A comparison of the fractional removal at steady state for PCP by sublation and by bubble fractionation on the small and large diameter columns is shown in Figure 6. The comparison was for identical volumetric gas and liquid flow rates on the two columns and hence the loading rates were not identical. It was observed that the effects of increasing column diameter on the fractional removal was greater in the case of bubble fractionation than in the case of solvent sublation. It is well known that the larger the column diameter, the larger is the re-dispersion of the solute at the top of the aqueous section in bubble fractionation(12). Wang and co-workers (12) also observed a precipitous drop in the fractional removal in bubble fractionation when the column diameter was increased. Similar observations for semi-batch bubble fractionation were reported by Shah and Lemlich(5). In solvent sublation the solute re-dispersion would be considerably reduced in solvent sublation because of the capacity of the organic solvent to capture and retain the solute carried up by the bubble(10). Liquid circulations in the top region of the aqueous section (so-called "exit region") in large diameter bubble columns arise from disturbances on the liquid surface set up by breaking bubbles. These disturbances are propagated through the liquid surface into the bulk liquid leading to redispersion of the solute carried by the bubbles. The presence of a thin film of organic solvent on the aqueous surface will eliminate such disturbances on the aqueous surface. Thus the liquid circulations leading to axial dispersion of the solute in a solvent sublation experiment should be considerably reduced as compared to a bubble fractionation experiment. These considerations may help explain the reduced dependence of column diameter on the fractional removal in solvent sublation as compared to bubble fractionation. The presence of organic solvents on the aqueous surface and its effects on axial dispersion of solutes in the aqueous phase is the focus of current investigations in our laboratory.

CONCLUSIONS

Both solvent sublation and bubble fractionation are viable as continuous countercurrent processes for the removal of hydrophobic compounds from water. Both processes are primarily dependent on the size of air bubbles introduced into the column as well as the extent of axial dispersion in the aqueous phase. The fractional removal in solvent sublation is less dependent on the column diameter.

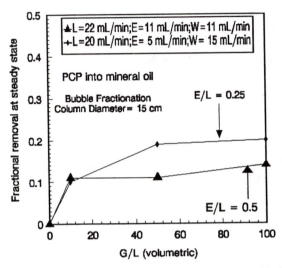

Figure 5. Effect of air/water volumetric flow ratio on the bubble fractionation of PCP on the large-scale bubble column.

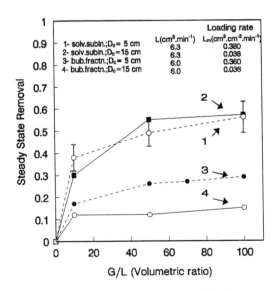

Figure 6. Comparison of the dependance of fractional removal in bubble fractionation and solvent sublation on the column diameter.

However, from the standpoint of scale up, both bubble fractionation and solvent sublation are severely limited by the low gas and liquid loading rates necessary for efficient separations. In comparison to other industrial wastewater treatment processes for hydrophobic compounds, both bubble fractionation and solvent sublation are at the present time unsatisfactory in terms of both the degrees of separation and the gas and liquid loading rates currently achievable in the bubble columns used for these processes. A lot more remains to be done before these techniques can find practical utility in large scale treatment processes.

Acknowledgment: This research was supported by a grant from the National Science Foundation (BCS-8822835).

LITERATURE CITED

1. *Adsorptive Bubble Separation Techniques*, Lemlich, R., Ed.,Academic Press: New York, NY, **1972**.
2. Clarke, A. N.; Wilson, D. J. *Foam Flotation: Theory and Applications*, Marcel Dekker, New York, NY, **1983**.
3. Lu, X. Y.; Valsaraj, K. T.; Thibodeaux, L. J. *Sep. Sci. Technol.* **1991**, *26*, 977-989.
4. Bruin, S.; Hudson, J. E.; Morgan, A. E. *Ind. Eng. Chem. Fundam.* **1972**, *11*, 175-181.
5. Shah, G. N.; Lemlich, R. *Ind. Eng. Chem. Fundam.* **1970**, *9*, 350-355.
6. Valsaraj, K. T.; Thibodeaux, L. J. *Sep. Sci. Technol.* **1991**, *26*, 37-58.
7. Valsaraj, K. T.; Thibodeaux, L. J. *Sep. Sci. Technol.* **1991**, *26*, 367-380.
8. Valdes Krieg, E.; King, C. J.; Sephton, H. H. *AIChE J.* **1975**, *21*, 400-402.
9. Rice, R. G.; Littlefield, M. A. *Chem. Eng. Sci.* **1987**, *42*, 2045-2053.
10. Valsaraj, K. T.; Lu, X. Y.; Thibodeaux, L. J. *Water Res.* ,**1991**, *25*, 1061-1072.
11. Wilson, D. J.; Pearson, D. E. *Solvent sublation of organic contaminants for water reclamation*, Report No: RU-84/5, U. S. Department of Interior, Washington, D. C. **1984**.
12. Kown, B. T.; Wang, L. K. *Sep. Sci.* **1971**, *6*, 537-552.

RECEIVED March 31, 1992

Chapter 10

Recovery of Valuable Metals from Industrial Wastes

S. Natansohn, W. J. Rourke[1], and W. C. Lai[2]

GTE Laboratories Incorporated, Waltham, MA 02254

Environmental and economic concerns necessitate the recovery of the heavy and valuable metals present in industrial processing wastes. Fundamental chemical principles were used in the development of a process for the separation and recovery of the variety of metals present in the residues from tungsten ore processing plants. Such ore tailings typically contain residual W and small quantities of Co, Cr, Nb, Ni, Pb, Ta, Th, Zn, lanthanides, and Sc in a matrix of Fe and Mn oxides. The emphasis of this separation path was the early, selective, and quantitative recovery of scandium which is present in concentrations of about 500 ppm. This was accomplished by acid leaching of the waste material in the presence of a reductant and passing the pH-adjusted leachate through a column of a complexing ion exchange resin. The column raffinate was then treated sequentially to separate and recover the other metals. Tungsten, Nb, and Ta were recovered from the solid leaching residue.

Waste products from many industrial processes contain at times significant concentrations of metals which are objectionable on environmental grounds and yet constitute an appreciable economic asset. The presence of toxic metals in such wastes constitutes an environmental hazard, particularly because the ever-decreasing pH of the rainwater makes their leachability and contamination of the ground water more likely. Careful and costly waste-disposal procedures are thus mandatory so as to prevent this from occurring. A preferred alternative is the cost-effective conversion of such waste into useful products. This provides an optimal solution to the waste-disposal problem because (a) it eliminates the need for a safe and costly disposal of the hazardous waste; (b) it maximizes resource utilization and conservation through recycling, and (c) it derives an economic benefit from the sale of the obtained products.

[1]Current address: Duracell Worldwide Technology Center, 37 A Street, Needham, MA 02194
[2]Current address: Aneptek Corporation, 209 West Central Street, Natick, MA 01760

0097–6156/92/0509–0129$06.00/0
© 1992 American Chemical Society

Consequently, the objective of this study was the development of an industrially viable, cost-effective, environmentally compatible technology for the recovery of metals from tungsten ore tailings. The methodology to be developed was to be sufficiently broad and flexible so as to be applicable to a wide range of metallic constituents present in varying proportions, and for this reason alone, tungsten ore tailings were a very appropriate test system. The following are some of the criteria deemed necessary for the success of such a metal-recovery process, i.e., the process needs to be:

• Waste reducing — the residue of the recovery process should be below 20% of its original mass and be acceptable to a sanitary landfill.

• Environmentally compatible — in that no hazardous reagents are used in the recovery process nor hazardous products generated.

• Industrially viable — the processing steps to be well-established and practiced unit operations.

• Cost effective — the overall processing cost should be significantly exceeded by the product value of the recovered constituents.

• Flexible — the process developed should be applicable to a broad range of compositions found in tungsten ore tailings.

• Generic — the technology should have general applicability to the recovery of metals from secondary sources such as low grade ores, ore tailings, scrap products, and industrial wastes, recognizing, however, that the specific chemical characteristics of each material define the most effective ways of its treatment and utilization.

Tungsten Ore Tailings

Origin. The tungsten ore tailings are the solid residue of the digestion process used for the recovery of tungsten from its ores. There are two economically important classes of tungsten ores: scheelite, $CaWO_4$, and wolframite, $(Fe_{1-x}Mn_x)WO_4$, which comprises a continuum of Fe and Mn concentrations from ferberite ($x = 0$) to huebnerite ($x = 1$). These naturally occurring minerals also contain minor amounts of other elements which fit into their structure either substitutionally or interstitially. The extraction of the tungsten values from wolframite ore concentrates is usually done at 100°C by reaction with strong solutions of NaOH as given by equation 1 (*1*):

$$(Fe,Mn)WO_{4(s)} + 2\ NaOH_{(aq)} \rightarrow (Fe,Mn)(OH)_{2(s)} + Na_2WO_{4(aq)} \tag{1}$$

The sodium tungstate is the process solution from which the tungsten is recovered by further treatment. The insoluble hydroxides are filtered off and constitute the tungsten ore tailings, the subject of this study. The tailings are usually stored in a self-contained site prior to ultimate disposal.

Composition. The tungsten ore tailings are a heterogeneous material of a complex and variable composition because they result from ores originating from different sources and having a broad range of stoichiometries. This product is a mixture of iron, manganese, calcium, sodium, and silicon oxides and hydroxides, and dispersed in this matrix are varying amounts of other elements. A typical, but not representative, composition obtained by mass spectrometric analysis which gives the concentration of the 20 most abundant elements is given in Table I. While this analytical method is not rigorously quantitative because of large errors, it does provide for a convenient detection of the elements present and an estimate of their relative concentration. A rigorous chemical analysis of this sample by gravimetric, spectrophotometric, and spectrometric techniques provided similar results. Thus, the concentration of Fe and Mn found in the sample was 23.2 wt % and 20.6 wt %, respectively, close to the values reported in Table I for these two most abundant constituents.

Table I. Typical Elemental Composition of the Tungsten Ore Tailings

Element	wt %	Element	wt %
Fe	22	Mg	0.21
Mn	20	Pb	0.19
W	4.3	Ni	0.15
Si	3.2	Co	0.11
Sn	0.96	Bi	0.10
Na	0.84	Ce	0.082
Ca	0.75	Th	0.067
Ti	0.40	Cu	0.062
Ta	0.38	K	0.061
Nb	0.30	Sc	0.050

These analytical experiments were used to establish the validity of the techniques used as well as the general concentration ranges of the metals of interest. However, because of the aforementioned material heterogeneity, it was necessary to determine the composition of each experimental sample in order to establish separation effects in relation to a particular initial composition. This is apparent in the subsequent data that report different base compositions. In spite of the wide variation in the ore tailings composition, commonalities in chemical behavior and reactivity which were used in the design of an effective separation and recovery process have been observed.

Analytical Techniques. The primary method used to determine the metallic element concentration in the tailings was DC Plasma Atomic Emission Spectrometry (DCP). It was used for the determination of both major and minor components. In the former case, the analysis is straightforward, but in the case of minor constituents, it was necessary to use "matrix matching," i.e., to use standard solutions having the same concentration of the major component as the unknown, to compensate for the background emission interference of the other solutes. This requires the initial determination of the major components to define the appropriate doping levels. The detection wavelengths used for the three elements of most interest were 259.940 nm for Fe; 259.373 nm for Mn; 391.181 nm for Sc at concentrations above 10 ppm, and the more sensitive 424.683 nm wavelength for Sc below 10 ppm. Under these conditions, it is possible to obtain reasonable measurement accuracy and precision, as shown by the data in Table II.

Table II. Concentration of Fe, Mn, and Sc in Tungsten Ore Tailings

Sample No.	Sample Weight (g)	Fe (mg/g)	Mn (mg/g)	Sc (mg/g)
1	9.8624	164.1	184.2	0.966
2	9.9228	163.2	178.0	0.884
3	9.9129	166.0	171.2	0.870
4	9.9302	164.5	186.4	0.874
5	9.9150	168.7	191.2	0.953
6	9.9994	173.0	191.4	1.004
Average		167 ± 4	184 ± 4	0.925 ± 0.056

In addition to the DCP determinations, other analytical techniques were employed as necessary. One of these was the colorimetric determination of ferric ion by the thiocyanate method (2). This was used to calculate the ferrous iron content which is the difference between the total iron, determined by DCP, and the trivalent iron obtained by the thiocyanate reaction.

Dissolution Studies

Environmental considerations, the chemical nature of the ore tailings, and the goal of recovering all metal components, including those present in trace amounts, dictate a hydrometallurgical rather than a pyrometallurgical approach to constituent recovery. Once the metals are in solution, subtle differences in their chemical behavior can be exploited to achieve separations which would not be possible by physical means. The emphasis of this study was on the early and complete separation and recovery of scandium from the matrix followed by further processing steps to recover the other components. In order to facilitate process design, it was necessary to establish the dissolution behavior of the tungsten ore tailings and the distribution of constituents within this material.

The results of the first dissolution experiments have shown that the tungsten ore tailings do not have a homogeneous composition but are a complex material consisting of a variety of constituents with different stoichiometry, structure, chemical reactivity, and solubility. Thus, the weight fractions reported in Tables I and II are "macro" representations of the overall sample content but do not imply that these constituent ratios can be found uniformly in every particle throughout its mass.

Acid Dissolution. In view of the fact that the oxides of two transition metals, Fe and Mn, account for 65% of the mass of the tungsten ore tailings, acids are the obvious choice for material dissolution. It was found that concentrated HCl dissolves up to 80% of the total mass, including nearly all of the Fe, Mn, and Sc content, leaving behind a residue which is quite different in nature. While the starting material is black, soft, and fairly reactive, the insoluble residue obtained from the acid digestion, the so-called "acid residue," is much lighter in color, sandy in texture, and unreactive. It contains primarily W, Si, Nb, Ta, and other acid-insoluble constituents.

A quantitative assessment of the solubility of Fe and Sc was made by treating a 5 g sample of the ore tailings with 6 N HCl. The insoluble residue amounted to 20.7 wt % of the original sample weight. This residue was fused with sodium carbonate and the fusion mass dissolved in 1 N HCl. The concentration of Fe and Sc was measured in both solutions, and the results are summarized in Table III. These data indicate that in spite of the large difference in the absolute quantities of Fe and Sc, the weight fraction of these metals which is soluble in 6 N HCl is about the same, in excess of 95%.

Table III. Distribution of Fe and Sc

	Present in Sample (mg)	(wt %)	Acid Soluble (mg)	(wt % of Total)	Acid Insoluble (mg)	(wt % of Total)
Fe	1106	22.1	1083	97.9	23.1	2.1
Sc	4.16	0.083	3.97	95.5	0.19	4.6

Particle Size Effects: A manifestation of the nonuniform nature of the tungsten ore tailings may be found in the results of the analysis of solutions stemming from digestion of various-size fractions of the material with concentrated HCl. The data are presented in Table IV.

Table IV. Acid-Soluble Fe and Mn in Tungsten Ore Tailings' Size Fractions

Size Fraction (Mesh)	Sample Weight (g)	Weight of Residue (g)	Fraction Dissolved (wt %)	[Fe] (mg/g)	[Mn] (mg/g)	[Fe]:[Mn]
+20	1.0533	0.3635	65.5	168.5	167.5	1.006
-20 +40	1.0177	0.2658	72.9	193.9	186.0	1.042
-40 +50	0.9423	0.2061	78.1			
-40 +50	1.0297	0.2292	77.7	219.6	201.0	1.093
-40 +50	1.0139	0.2215	78.2			
-50 +150	1.0747	0.2177	79.7	224.3	203.3	1.103
-150	1.0437	0.2334	77.6	218.4	185.2	1.179
Overall						1.13

While it is to be expected that the larger particles will be more difficult to dissolve because of the smaller specific surface area, the observed differences indicate that the larger size fractions contain a signifigantly larger portion of acid-insoluble components, particularly in view of the strong acid treatment of this experiment. The acid-soluble portion amounts to about 78% of the sample weight; the reproducibility of this value is good, as indicated by the precision of the measurement performed on three different samples of the $-40 +50$ size fraction. The interesting observation is the variation in the relative amounts of Fe and Mn found in the different size fractions. The ratio of Fe to Mn increases with decreasing particle size of the samples, indicating that the finer particles are relatively richer in Fe, while the coarser ones have a higher Mn content. The overall ratio of Fe:Mn in this sample as obtained from complete sample dissolution is 1.13. These data show that a substantial segregation exists between the two major constituents. The two metals appear to be present in more than just a single mixed metal compound, but rather in several discrete metal compound phases.

Reaction Mechanism

While the alkaline digestion of wolframite ores results in the formation of $Mn(OH)_2$ (equation 1) in which Mn is divalent, the presence of MnO_2 in the ore tailings is not surprising because the air oxidation of $Mn(OH)_2$ is well known (3). Extensive observations of the dissolution behavior of the tungsten ore tailings indicate that it is not a simple acid dissolution but a redox reaction such as

$$MnO_{2(s)} + 4\ HCl_{(aq)} \rightarrow Mn^{2+}_{(aq)} + 2\ H_2O + 2\ Cl^-_{(aq)} + Cl_{2(g)}. \qquad (2)$$

Tetravalent manganese is not soluble in acids; it must be reduced to the divalent state in order to dissolve. Hydrochloric acid is a useful reductant for this purpose, and in the process, a stoichiometric amount of chloride is oxidized to chlorine gas. The evolution of this gas is indicative of the redox nature of the reaction.

In order to confirm the redox nature of the dissolution reaction and to determine the amount of tetravalent Mn present, the dissolution was carried out in a closed system, and the evolved gas, carried in an argon stream, collected in a KI solution. Chlorine oxidizes iodide to iodine, which can then be titrated by a solution of sodium thiosulfate in a classic iodometric reaction (4). This experiment was performed using a sample of the ore tailings and also with pure reagent grade MnO_2, which was used for calibration purposes. The results indicate that about 78% of the total Mn is tetravalent. Similar results were obtained from the reduction of MnO_2 by oxalic acid in a sulfuric acid solution (5). These results lead to the conclusion that the acid-soluble portion of the material consists of a fraction, primarily MnO_2, which requires reduction to become soluble, and one composed of other Mn/Fe oxides that is directly acid soluble.

Dissolution System. The effectiveness of HCl in dissolving the tungsten ore tailings is based on the fact that it is a reducing agent for MnO_2. However, the evolution of chlorine (equation 2) renders the process objectionable for industrial applications because of environmental, safety, and health concerns. The key conclusion of the dissolution studies is that it is necessary to reduce the MnO_2 fraction of the ore tailings in order to bring essentially all of the Mn and Sc into solution. Alternate reagents need to contain a reductant which is best dissolved in a nonreducing acid such as H_2SO_4. After an evaluation of several reductants, such as hydrogen peroxide, hydroxylamine, hydroquinone, oxalic acid, and hydrazine, it was concluded that a solution of the latter in sulfuric acid was best suited for the purpose. The reduction of manganese dioxide by hydrazine proceeds according to equation 3:

$$2 MnO_2 + N_2H_4 + 4 H^+ \rightarrow 2 Mn^{2+} + N_2 + 4 H_2O \qquad (3)$$

The reduction of Mn^{4+} by hydrazine in an acid medium is favored kinetically and, as expressed in equation 3, consumes hydrogen ions. Consequently, there is initially no significant decrease in the pH of the system despite rather large additions of sulfuric acid. Only when all of the tetravalent manganese has been reduced, as indicated by a constant concentration of Mn in the solution, does the pH begin to decrease with further increases in acid concentration. This is shown in Figure 1, in which the pH of the solution and its Mn concentration are plotted as a function of the normality of sulfuric acid at constant hydrazine concentration (each data point represents an individual sample). The hydrazine/sulfuric acid reductant used consisted of 0.132 M $N_2H_2 \cdot H_2O$ dissolved in 0.625 M H_2SO_4; higher concentrations may cause the crystallization of hydrazine sulfate, which is difficult to redissolve. The two reagents were diluted prior to mixing and then stirred vigorously. The tungsten ore tailings were digested at room temperature in the amount of 50 g per liter of this reagent.

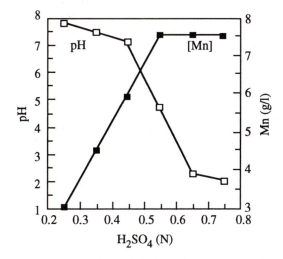

Figure 1. Mn concentration and pH as a function of acid concentration in sulfuric acid/ hydrazine systems.

In another experiment, the dissolution characteristics of the tungsten ore tailings in this reagent system were examined by digesting six samples in solutions of varying acid but fixed hydrazine concentration. The reagent mixtures were filtered and the filtrates analyzed for Fe, Mn, and Sc. The filtration residues were dried and weighed to determine the total amount of sample dissolved. Then the residues were dissolved completely in hot HCl and the amount of Fe, Mn, and Sc determined. These data were used to calculate the amount of each metal dissolved at a particular acid concentration. The results are depicted in Figure 2.

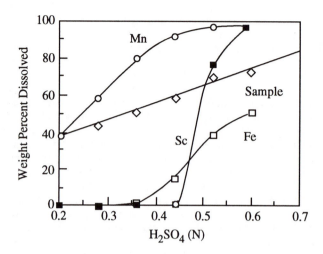

Figure 2. Dissolution of ore tailings as a function of sulfuric acid concentration.

The results of this experiment indicate that the total amount of sample dissolved is linearly dependent on the concentration of sulfuric acid, but that the dissolution of the key elements, Fe, Mn, and Sc, does not proceed in a parallel fashion. The dissolution of Mn proceeds rapidly and preferentially consistent with the postulated reduction mechanism and confirming the data in Figure 1, which were obtained on a different set of samples. The next constituent to be dissolved is scandium, which is being solubilized upon completion of the tetravalent Mn reduction and the concomitant decrease in pH; the Fe dissolution proceeds at a slower rate. These data indicate that there is a possibility of isolating in the pH range of 7 to 8, where the dissolution of Mn is dominant, a relatively pure solution containing up to 80% of the Mn values with a Fe concentration of less than 2000 ppm. This is remarkable considering that this separation occurs in a single step from a system in which both metals are present in the solid matrix in roughly equal concentrations of about 20 wt % each.

The key trends observed in the dissolution of tungsten ore tailings are independent of the reductant system or sample variation. These are the following:

- The linear dependence of the material dissolution on acid concentration at fixed reductant concentration.

- The selective dissolution of Mn with initially no attendant Fe or Sc dissolution.

- The initiation of the Fe and Sc dissolution only after most of the Mn has been reduced and dissolved.

- The rapid dissolution of Sc, which is completed while only about 50% of the Fe is dissolved.

The experimental evidence obtained indicates that the three key constituents may be segregated within the tungsten ore tailings whose acid-soluble fraction consists of three major phases:

- A MnO_2 phase which needs to be reduced to render it acid-soluble, contains up to 80% of the total Mn and which may be selectively dissolved with the virtual exclusion of Fe and Sc.

- An acid-soluble Mn/Fe phase which contains 90% to 95% of the Sc and about 50% of the total Fe.

- A relatively less reactive Fe phase that consists substantially of magnetite, Fe_3O_4, as indicated by magnetic measurements and which contains little Mn or Sc.

This knowledge of the composition and nature of the starting material, particularly the distribution of the constituents throughout its mass, is essential to the design of an effective recovery process (6).

Metal Separation and Recovery Scheme

The objective of this effort was the initial recovery of Sc with subsequent process steps to separate and recover the other metallic constituents. The initial step was the acid digestion of the tungsten ore tailings, which produced an insoluble residue containing W and a solution rich in most of the other metals. The steps in the overall separation scheme are outlined by the flow sheet presented in Figure 3. The hydrazine/ sulfuric acid solution was chosen as the reductant/solvent reagent based on the following considerations:

- Effectiveness: Both MnO_2 and Fe^{+3} are reduced by the reagent. The latter is important for the ion exchange recovery of Sc^{+3}, which does not occur in the presence of ferric ion.

- Ease of operation: The system is dilute, which reduces process equipment corrosion, and operates at room temperature, which saves energy; the digestion product is readily filterable.

- Convenience: The chemicals are readily available, and the reaction products are innocuous. The end pH of the reaction is about 2, which is suitable for the Sc extraction, thus requiring no, or a minimum, pH adjustment.

- Flexibility: The option exists of separating a pure Mn process stream up front.

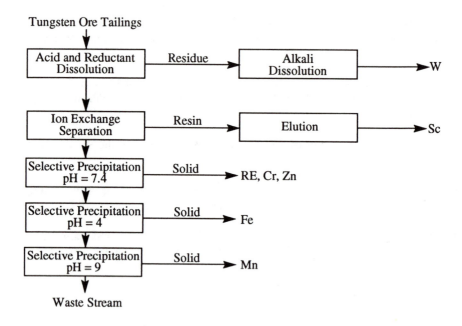

Figure 3. Metal separation and recovery path.

Extraction and Separation of Scandium. The dissolution of the tungsten ore tailings results in a solution with a small amount of Sc, typically about 0.03 g/l, in the presence of a large excess of Mn and Fe, for example 8.0 and 4.9 g/l, respectively. The goal of an early, selective, and quantitative separation of Sc from this process stream makes a liquid-liquid extraction technique, such as occassionally described for this process (7), impractical. Liquid-liquid extraction is not a preferred unit operation because it generally consists of several steps, there are problems with organic extractant and solvent entrapment, and, in the case in hand, the neat separation of a small amount of Sc from a stream containing several hundred times larger Mn and Fe concentration would be very difficult. Extraction by an ion exchange resin appears more suitable for this purpose and offers the possibility, through the use of multiple parallel columns, of a continuous process.

The ion exchange resin selected for this study was a copolymer of styrene with divinylbenzene with a weakly basic iminodiacetic functional group. The reason for this choice is that it has been studied in the extraction of transition metal and other ions (8–10), is commercially available, and is being used in industrial applications. At first it would appear that the prevalent complexation of iminodiacetic acid with most metallic elements would preclude the type of selectivity sought for the Sc extraction. But such separations are possible by the exploitation of specific chemical behavior and complexation characteristics.

The structure of iminodiacetic acid is very similar to that of ethylene diamine tetraacetic acid (EDTA), it may be considered as one half of EDTA, and inasmuch as the latter's complexation behavior is well known and understood, it was used to model predictively the system under study. EDTA forms monodentate complexes with many metals, including Sc, and the extent of the complexation reaction is given by the stability constant, which varies for various metals over several orders of magnitude. The conditional stability constant treatment of Ringbom (*11*) allows simple predictions about the effect of pH, cation hydrolysis, and other ligands on the stability constant. The effect of pH is very pronounced; the absolute values of the stability constants decrease with decreasing pH, but the differences in magnitude between complexes of different cations are not reduced. The consequence of this is that at low enough pH, some metals do not form complexes while others do, that is, selective complexation can be achieved. This approach is demonstrated by the data in Figure 4, which identifies the minimum pH values for effective titration of various metal ions by EDTA (*12*). It is possible to selectively titrate one ion in the presence of another by choosing the appropriate pH of the solution and therefore the complex stability region.

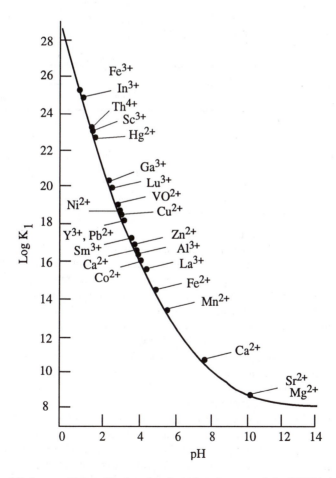

Figure 4. Minimum pH for effective titration of various metals by EDTA (reproduced with permission from reference 12; Copyright 1958 American Chemical Society).

The retention of the Sc by the ion exchange resin may be estimated by approximating the distribution coefficient of the metal between the resin and the aqueous phase as the conditional stability constant, $K_{ML'}$

$$D_{ML} = [M]_r/[M]_a \approx K_{ML'} \qquad (4)$$

which is obtained by dividing the stability constant by the alpha coefficient, a measure of the extent of side reactions (10). In the case of EDTA, where the dominant side reactions are those of the ligand, the logarithmic conditional stability constant is

$$\log K_{ML'} = \log K_{ML} - \log a_L. \qquad (5)$$

An examination of the data in Figure 4 indicates that a pH of about 2 should be appropriate for Sc complexation. An estimate of the logarithmic values of $K_{ML'}$ at pH 2 for EDTA complexes of the elements of interest is given in Table V.

Table V. Logarithmic Values of Stability Constants

Ion	K_{ML} (13)	α_L (14)	$K_{ML'}$
Sc^{+3}	23.1	13.7	9.4
Fe^{+3}	25.1	13.7	11.4
Fe^{+2}	14.3	13.7	0.6
Mn^{+2}	14.0	13.7	0.3
Th^{+4}	23.2	13.7	9.5
Ca^{+2}	10.7	13.7	< 0

The data indicate that, at a pH of 2, divalent Mn and Fe have much lower conditional stability constants than Sc, and therefore Sc would be complexed preferentially, even in the presence of larger concentrations of these ions. However, the $K_{ML'}$ of the ferric ion is larger than that of Sc, and it is thus mandatory to keep all iron in the solution in the ferrous state in order to achieve selective Sc complexation and, if this model is reasonable, selective retention on the ion exchange column. The data for Th and Ca are presented to illustrate the effect of other ions; Ca presents no problem as it is not complexed at this pH with EDTA, while Th would be co-extracted at a comparable rate with Sc. The assumptions of this model have been proven valid, as shown by the data below (Table VI) which demonstrate the selective extraction of Sc on the ion exchange column with iminodiacetic acid functionality in the presence of a large excess of Mn and Fe.

The scandium recovery process is outlined schematically in Figure 5. The tungsten ore tailings are dissolved in the hydrazine/sulfuric reagent, the pH is adjusted to 2.0, and the resulting solution is checked for the presence of ferric ion colorimetrically using the thiocyanate spot test. Any ferric ion content is reduced to the ferrous state by the addition of elemental iron as per the following equation:

$$Fe^0 + 2Fe^{3+} \rightarrow 3Fe^{2+} \qquad (6)$$

The solution, the typical composition of which is given in Figure 5, is then passed through a bed of an ion exchange resin with iminodiacetic functionality (which has been converted from its commercial sodium form to the hydrogen form by treatment with dilute sulfuric acid) at a rate of about 0.028 bed volumes (BV) per minute. The retention of the Sc is quantitative at first and begins to decrease gradually as the complexation sites on the resin become occupied (Table VI).

The concentration of Fe and Mn in the column effluent (Figure 5) is virtually the same, within the experimental error of the analytical technique, as that of the feed solution demonstrating that Sc is retained selectively in the presence of a 100- to 200-fold excess of these elements. In the experiment described, the Sc retention efficiency decreased below 90% after about 45 BV of solution has been passed and the experiment was concluded after 52 BV (Table VI). About 95% of the Sc present in the feed solution were retained on the column.

Table VI. Efficiency of Sc Retention on Ion Exchange Column

Effluent Fraction No.	Fraction Volume (ml)	Cumulative (ml)	Feed Volume (BV)	Sc Concentration in Fraction (ppm)	Sc Retention Efficiency (%)
1	1000	1000	7.7	0	100
2	3100	4100	31.5	0.22	99.2
3	1000	5100	39.2	0.88	96.9
4–11	400	5000	42.3	1.94	93.2
13–15	300	6800	52.3	7.85	72.7

The column was then washed with a solution of pH 2.0 to remove feed stream residues present in the interstitial volume of the column. The pH was chosen to maintain retention of the Sc on the column during this step and to prevent the complexation of other ions. The Sc was then eluted from the column quantitatively using 0.05 M diglycolic acid. The use of a complexing agent as an eluant affords an additional opportunity to improve the selectivity of the process. Diglycolic acid was selected as an eluant based on considerations similar to those discussed in the selection of the agent and conditions for the selective extraction of Sc. At the particular pH of the elution, the stability constants of diglycolic acid complexes with Sc are much larger than those of Sc with the iminodiacetate ligand of the ion exchange resin and also much larger than those of complexes between diglycolic acid and other column-retained ions of similar reactivity, such as thorium. The scandium is thus eluted preferentially and completely, and in the process, a purification step occurs because the other elements retained on the column are not eluted along with the scandium.

The scandium values may be directly recovered from the eluate by addition of ammonium hydroxide and consequent precipitation of scandium hydroxide at pH above 7. A preferred technique is the precipitation of scandium oxalate; it results in a purer product because many of the metallic impurities which co-precipitate as the

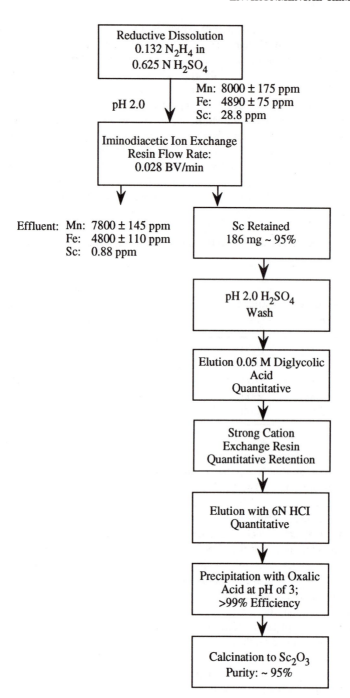

Figure 5. Scandium recovery scheme.

hydroxides with the scandium do not precipitate as the oxalates. However, the direct precipitation of scandium oxalate is inhibited in the diglycolic acid eluate because the scandium is complexed with the diglycolate ligand. A quantitative separation of Sc is effected by passing the eluate through a column of a strong cation ion exchange resin in the hydrogen form. As the Sc ions are retained on this ion exchange resin, the replaced H^+ ions protonate the diglycolate anions so that the column effluent is a regenerated diglycolic acid solution which can be used again in eluting a scandium loaded extraction column.

The scandium retained on the strong cationic resin is eluted quantitatively with 6 N HCl. In the process, the resin is regenerated by the strong acid and can be used again at undiminished capacity. The scandium is then precipitated by the addition of saturated oxalic acid. The precipitate is filtered, washed, and converted to Sc_2O_3 by calcination at elevated temperatures. The scandium recovery process is very efficient in that 99% of the Sc retained on the initial extraction column is recovered in the sesquioxide product. The overall recovery of the Sc present in the initial feed solution is determined solely by the point at which the extraction process is stopped.

The purity of the Sc_2O_3 is about 95%; the impurities are listed in Table VII.

Table VII. Impurity Content of Recovered Scandium Oxide

Element	wt %	Element	wt %
Ce	1.4	Pr	0.17
Th	1.0	Gd	0.16
Pb	0.85	Ca	0.16
Nd	0.53	Fe	0.077
La	0.27	Mn	0.044
All others	0.62		

The major impurities are those whose complexation characteristics are similar to scandium's and would be expected to be carried along in the process. The total of the unlisted impurities is 0.62 wt %, with none of them exceeding 0.1 wt %. The selectivity of the process is demonstrated by the fact that the scandium oxide contains only 0.07 and 0.044 wt % of Fe and Mn, respectively, whereas the starting material had 22 wt % Fe, 20 wt % Mn, and only 0.05 wt % of Sc.

Recovery of Metals Concentrate. The Sc-depleted raffinate from the ion exchange process step contains the two major constituents, Fe and Mn, in their divalent state, and other transition and rare metals in small amounts. The recovery of these metals in the presence of large amounts of Fe and Mn is done effectively by selective precipitation in the pH range between 6.5 and 7.5. In this experiment, the pH of the raffinate solution was adjusted with ammonium hydroxide to 7.4, and the resulting precipitate washed and dried. It contains the metals listed in Table VIII in a matrix of hydrated ferric oxide; the precipitation of appreciable amounts of iron, about 14% of the iron content of the raffinate, is primarily due to the partial oxidation of the ferrous to ferric ions during the ion exchange and selective precipitation steps. The concentration of these metals represents a useful upgrading over that in the original tungsten ore tailings and renders the precipitate a viable resource, particularly of the lanthanides.

Table VIII. Composition of Metal Concentrate

Element	wt %	Element	wt %
Si	5.9	Al	0.57
Mn	1.9	Co	0.25
Cr	1.0	Pb	0.23
Zn	0.91	Ca	0.064
Lanthanides	1.8	Ni	0.043
(incl. Y & Sc)			

Recovery of Iron. The Fe values are recovered from the filtrate of the preceding process step by oxidative precipitation to yield an easily filterable goethite solid, FeOOH. Filterability is a major technical issue in the removal of iron from industrial streams because of the difficulties encountered in filtering the gelatinous iron hydroxide precipitates. The solution was treated with hydrogen peroxide to oxidize the ferrous to ferric ions, the pH was adjusted to 4, and then digested at 85°C for about 3 hours to precipitate the goethite. The Fe recovery is virtually quantitative, >99%, yielding a high-purity product (Table IX).

Table IX. Impurities in Fe Precipitate

Element	w/o	Element	w/o
Mn	0.58	Ni	0.0026
Si	0.041	Na	0.0024
Zn	0.020	Cr	0.0016
Al	0.016	Ca	0.0015
Co	0.014	Cu	0.0009

Recovery of Manganese. The Mn content can be recovered from the filtrate of the goethite precipitation by another oxidative precipitation. The pH of the solution is brought to a value of about 9 by addition of NaOH, and an oxidant such as hydrogen peroxide is added to oxidize all of the divalent Mn. The resulting manganese hydroxide is easily filterable, contains in excess of 99% of the Mn present in solution, and its purity also exceeds 99% (Table X).

Table X. Impurities in Manganese Hydroxide

Element	wt %	Element	wt %
Co	0.34	Ni	0.023
Fe	0.16	Na	0.013
Si	0.069	Ce	0.012
Ca	0.032	Mg	0.011
Zn	0.026	La	0.0063

Composition of Waste Stream. The filtrate from the Mn precipitation is the waste stream resulting from this process. It has a pH of 9 and solid content of 17 g/l of Na_2SO_4. Its impurity content is given in Table XI, which lists the 10 highest contaminants. The concentrations are given in ppm; no other impurity was detected at a level greater than 50 ppb. The impurity content of this waste stream is quite low and, if not directly disposable, compatible with standard waste water treatment procedures. The low concentration of the transition metal ions, particularly Fe and Mn, which were major constituents of the feed stream attests to the effectiveness of the separation and recovery processes.

Table XI. Impurities Content of Waste Stream

Element	ppm	Element	ppm
Ca	58	Cl	0.7
Mg	20	Fe	0.5
Si	11	P	0.54
K	8.5	Al	0.30
Mn	3.3	Sr	0.17

Recovery of Tungsten. The residue from the initial acid reductant acid treatment of the tungsten ore tailings amounts to about 20% of their weight and contains close to half of the Fe content as well as the major fraction of W, Nb, and Ta. The Fe content is recoverable by treatment of this material with concentrated HCl at 85°C to obtain a liquid iron chloride concentrate which is separated by filtration. The insoluble fraction is digested in 6 N NaOH at 80°C to solubilize the tungsten. The resulting solution has a W concentration of 45 g/l, about 98% of the digested material's W content. The residue of this alkali treatment contains the Ti, Nb, and Ta in a silicate matrix.

Summary

The preceding discussion describes a sequential methodology for the separation and recovery of various metallic constituents from industrial wastes as exemplified by tungsten ore tailings. It was demonstrated that by exploitation of subtle differences in chemical behavior, it is possible to preferentially dissolve the constituents of the material, separate them selectively, and recover them quantitatively in materials of high purity. In the process, a potentially hazardous industrial waste is converted into useful products, virtually eliminating the disposal requirements and recycling strategic resources, while the effluent of the process contains only traces of the original components. The principles of this technology are applicable to many similar systems.

Literature Cited

(1) Mullendone, J.A. *Kirk-Othmer Encyclopedia of Chemical Technology*; 3rd Edition; John Wiley & Sons: New York, NY, 1983; Vol. 23, p. 417.

(2) Kolthoff, I.M.; Sandell, E.B. *Textbook of Quantitative Inorganic Analysis;* The Macmillan Company: New York, NY, 1948; pp. 672–673.

(3) Cotton, F.A.; Wilkinson, G. *Advanced Inorganic Chemistry;* Wiley Interscience: New York, NY, 1962; p. 695.

(4) Kolthoff, I.M.; Sandell, E.B. *loc. cit.,* pp. 614-619.

(5) *Ibid;* pp. 592-604.

(6) Welch, A.J.E. *Extraction and Refining of the Rarer Metals;* Institution of Mining and Metallurgy, London, 1957, p. 3.

(7) Guo Gongyi; Chen Yuli; Li Yu. *J. Met.* **1988**, *40* [7], 28.

(8) El-Sweify, F.H.; Shabana, R.; Abdel-Rahman, N.; Aly, H F. *Solvent Extraction and Ion Exchange*, **1986**, *4*, 599.

(9) Hubicki, Z. *Hydrometallurgy,* **1986**, *16*, 361.

(10) Diaz, M.; Mijangos, F. *J. Met.* **1987**, *39* [7], 42.

(11) Ringbom, A.J. *Complexation in Analytical Chemistry;* Interscience Publishers; New York, NY, 1963; pp. 22-60.

(12) Reilley, C.N.; Schmid, R.W. *Anal. Chem.* **1958**, *30*, 947.

(13) Ringbom, A.J. *loc. cit.,* pp. 332-333.

(14) *Ibid,* p. 351.

RECEIVED April 7, 1992

Chapter 11

Removing Heavy Metals from Phosphoric Acid and Phosphate Fluid Fertilizers

Organic and Inorganic Reagents

V. M. Norwood III[1] and L. R. Tate

Chemical Research Department, Tennessee Valley Authority,
National Fertilizer and Environmental Research Center,
Muscle Shoals, AL 35660–1010

Heavy metals in wet-process phosphoric acid (WPA) and phosphate fluid fertilizers are an environmental concern in the United States and Europe. To address this concern, several organic and inorganic reagents were evaluated as precipitants for heavy metals in a 10-34-0 (N-P_2O_5-K_2O) fluid fertilizer and WPA. Trisodium trithiocyanuric acid (TMT-15), sodium polythiocarbonate (Thio-Red II), and sodium trithiocarbonate (5% Na_2CS_3) precipitated arsenic, cadmium, copper, mercury, lead, and zinc from 10-34-0. Ammonium cyanurate was ineffective in removing cadmium from 10-34-0. Thio-Red II and 5% Na_2CS_3 precipitated mercury, lead, cadmium, copper, and chromium from WPA. A water-insoluble starch xanthate adsorbed mercury, copper, and lead from 10-34-0 and WPA. Sodium sulfide, sodium polysulfide, and potassium ferrocyanide were tested as inorganic precipitants. The polysulfide was twice as effective as the sulfide alone, and concentrations of less than 10 ppm of arsenic, cadmium, mercury, and lead were achieved in 10-34-0. Ferrocyanide reduced the concentrations of cadmium and nickel to less than 10 ppm in WPA.

Phosphate fluid fertilizers of various grades [e.g., 10-34-0 (N-P_2O_5-K_2O)] are prepared by ammoniating merchant-grade superphosphoric acid. The superphosphoric acid is produced by concentrating merchant-grade orthophosphoric acid (54% P_2O_5) in a vacuum or atmospheric concentrator. Similarly, merchant-grade orthophosphoric acid is produced by concentrating wet-process phosphoric acid (WPA, 30% P_2O_5). WPA is manufactured by acidulating the apatite content of phosphate rock with sulfuric acid to produce phosphoric acid and insoluble calcium sulfate dihydrate (gypsum).

The majority of the metallic impurities originally present in the phosphate rock, such as iron, aluminum, magnesium, zinc, lead, cadmium, manganese, nickel, chromium, vanadium, mercury, and arsenic, are acid-soluble, and these become partitioned between the WPA (80%) and the solid waste gypsum (20%) during the acidulation step. Consequently, some portion of these metallic impurities also is found in phosphate fluid

[1]Current address: 1101 Market Street, MR–3A, Chattanooga, TN 37402

fertilizers. Table I summarizes the levels of heavy metal impurities found in phosphoric acids and 10-34-0 phosphate fluid fertilizers manufactured in Florida, North Carolina, and the Western states (Wakefield, Z.T., TVA Bulletin Y-159, Tennessee Valley Authority, unpublished data).

Table I. Concentrations of Heavy Metals in Phosphoric Acids and Phosphate Fertilizers

Source	P_2O_5 wt %	Heavy Metals (ppm) Cd	Cu	Zn	Mn	Ni	Pb	Cr	V
Central Florida									
Acid	51.6	6	12	70	346	36	5	86	212
10-34-0	34.0	4	8	46	228	24	3	57	140
North Florida									
Acid	47.8	8	15	98	371	13	1	82	112
Superacid	69.3	11	19	149	610	25	<1	126	161
10-34-0	34.0	5	9	73	299	12	<1	62	79
North Carolina									
Acid	51.2	49	2	500	88	37	<1	256	55
Superacid	68.8	59	<1	700	123	51	<1	345	72
10-34-0	34.0	29	<1	346	61	25	<1	170	36
Western Region									
Acid	53.4	128	55	1202	119	107	1	632	876
Superacid	68.7	113	<1	1651	145	194	7	780	1474
10-34-0	34.0	56	<1	817	72	96	3	386	729

As part of its environmental initiative, the Tennessee Valley Authority's (TVA) National Fertilizer and Environmental Research Center has redirected the objectives of some chemical research projects to address environmentally-related concerns of the fertilizer industry in the United States (1). In the United States, California Proposition legislation could place severe limits on the maximum concentrations of cadmium, lead, arsenic, mercury, and chromium allowed in WPA and phosphate fluid fertilizers. Legislative agencies in some European countries already have implemented maximum allowable concentrations of heavy metals in phosphate fertilizer products. For example, in April 1987 the European Economic Community (EEC) published its proposals for dealing with the matter of heavy metal pollution, specifically cadmium, resulting from the use of phosphate fertilizer products (2-5). The heavy metals mercury, lead, cadmium, and arsenic present a particularly severe environmental problem since they can only be isolated by techniques such as precipitation, adsorption, solvent extraction, or physiochemical reduction processes down to the elements themselves. Most of the schemes for removing heavy metals from WPA before it is converted into phosphate fertilizers are based on solvent extraction (*6-8*). Few studies have been reported on the use of precipitants or adsorbents to remove heavy metals from WPA or phosphate fluid fertilizers. Other schemes for removing metals from WPA and phosphate rock include ion exchange (*9*) and calcination in a fluidized bed, respectively.

Organic Reagents for Heavy Metal Removal

Thiocarbonate Precipitants. The use of organic precipitants, such as poly-thiocarbonates and dithiocarbonates, as an effective technique for removing heavy metals from waste water was identified during a computerized literature search. These organic precipitants, when added to a waste stream, form a dense stable floc that is neither dependent on pH nor affected by chelated or other complexed metals. The sludges formed by polythiocarbonate reagents, for example, are both stable and capable of passing most Environmental Protection Agency (EPA) leachate testing. Heavy metals can also be precipitated from waste water using dithiocarbamates (*10*); however, the fungicidal effect of this class of reagents has restricted their use. Consequently, the ability of dithiocarbamates to precipitate heavy metals from either 10-34-0 or WPA was not investigated in this study.

Elfine (*11*) reported that 5% Na_2CS_3 was highly effective in precipitating iron(III) and chromium(III) from actual industrial waste water samples. For example, an effluent from a galvanizing plant (pH adjusted to 6-9.5) containing 523 ppm iron was treated with the trithiocarbonate; and after a 5 min reaction time and subsequent workup, the effluent was found to contain only 0.23 ppm iron.

Starch Xanthate Adsorbents. Wing and coworkers (*12*) reported that insoluble starch xanthate was highly effective in adsorbing iron(II) and chromium(III) from synthetic waste water samples. For example, a pH 3.0 waste water with an initial iron(II) concentration of 27,920 mg/L was treated with the starch xanthate; and after stirring for 2 h, the residual iron(II) concentration was found to be <1 mg/L. Similar results were obtained for chromium(III), where the chromium(III) concentration of a synthetic waste water sample at pH 3.2 was reduced from 26,000 mg/L to 3 mg/L after stirring for 2 h with the appropriate amount of starch xanthate.

Inorganic Reagents for Heavy Metal Removal

Sulfide Precipitants. The removal of heavy metals from waste waters using soluble sulfides has been studied extensively. Peters and Ku (*13*) showed that pH strongly affects the removal of heavy metal sulfides from waste water. Their data showed that the solubilities of most heavy metal sulfides, other than arsenic (III) sulfide, decreases up to about pH 9. Since most phosphate fluid fertilizers, such as 10-34-0 and 11-37-0, have a pH in the range of 6 to 7, these data indicate that concentrations of several heavy metals in phosphate fluid fertilizers could be reduced to very low levels by precipitation with inorganic sulfide reagents. The use of soluble sulfides to precipitate heavy metals from WPA has also been reported. Both Maruyama (*14*) and Berglund (*15*) have demonstrated that arsenic concentrations in WPA could be reduced to <1 ppm under certain conditions.

Although sulfides have been studied as potential inorganic reagents for removal of heavy metals from various waste waters and WPA, very little work has been done with the polysulfides. Solutions of polysulfides are easily prepared by heating a solution of a metal sulfide with elemental sulfur. Pickering and Tobolsky *(16)* state that polysulfide solutions reach equilibrium very quickly. As a matter of convenience, the sodium polysulfides may be given the general formula Na_2S_n, where $\underline{n}$ can range from 1 (the normal sulfide) to 5 (the pentasulfide).

Ferrocyanide. The reaction of Fe(II) and Fe(III) with the ferrocyanide ion $Fe(CN)_6^{4-}$ to form highly insoluble compounds is well known. However, cadmium, lead, and probably other heavy metals also form ferrocyanide precipitates which may allow their removal from WPA and phosphate fluid fertilizers.

Potassium ferrocyanide was tested in the 1960s at TVA as a means for removing iron from WPA (Wakefield, Z.T.; McCullough, J.F.; Wright, B., Tennessee Valley Authority, unpublished data). The precipitate was noted to be "Prussian Blue", which has the general composition $MFe^{3+}[Fe(CN)_6]$ where M is a monovalent ion, normally K^+. These TVA reports also described a method of recovering the ferrocyanide for recycle.

Objective

The objective of this project, performed in 1987 and 1988, involved reduction of metallic impurities, specifically heavy metals, in phosphate fluid fertilizers and WPA. If fertilizer materials do not continue to receive an exemption under the California Proposition legislation, then both WPA and phosphate fertilizer manufacturers will need a method to reduce the concentrations of heavy metals in their products. Since few studies have been reported recently where the use of precipitating or adsorbing agents to remove heavy metals from WPA or phosphate fluid fertilizers has been investigated, organic and inorganic reagents which effectively precipitate or adsorb heavy metals from waste water were identified in this study and tested for their ability to remove heavy metals from 10-34-0 and WPA. The 10-34-0 phosphate fluid fertilizer was selected for study because this material supplies over 50% of the base solution used for production of other phosphate fluid fertilizers with typical grades of 7-21-7, 4-10-10, 6-18-6, and 20-10-0 *(17)*.

Experimental

Synthesis of Organic Reagents. The literature methods were used to synthesize the 5% Na_2CS_3 *(18)* and the water-insoluble (WI) starch xanthate *(12)*. For the preparation of the 5% Na_2CS_3 solution, an aqueous solution of sodium hydroxide was reacted at room temperature for 16 h with carbon disulfide. The orange-red solution obtained from this reaction was then diluted with water to give a 5% solution. In a typical preparation of the WI starch xanthate, 35.4 g of a highly crosslinked starch (53-91E, Hubinger Co.) was slurried in water (225 mL), 8 g sodium hydroxide in water (100 mL) was added, and the mixture was stirred 30 min. Carbon disulfide (5 mL) was then added and the mixture stirred 16 h at $25^{\circ}C$. The resulting slurry was filtered and the solid washed successively with water, acetone, and ether. After drying in a vacuum oven at $25^{\circ}C$ for 2 h, the pale-yellow WI starch xanthate (%S, 5.5; %H$_2$O, 12.3) was stored at $0^{\circ}C$ in a closed container.

TMT-15 (Degussa Corp.) and Thio-Red II (Environmental Technology, Inc.) were obtained from commercial sources. All heavy metal salts (CuSO$_4$, As$_2$O$_3$, MnCl$_2$·4H$_2$O, HgCl$_2$, CdCO$_3$, PbCO$_3$) and K$_4$Fe(CN)$_6$ were reagent grade (98+ % purity) and used as received. Some of the heavy metals were present in the 10-34-0 and WPA in low concentrations; therefore, additional heavy metal was added as required to reach a concentration of 20-150 ppm in each case. Ammonium cyanurate was prepared by adding ammonium hydroxide solution to a mixture of cyanuric acid and water at pH 7-8, heating the mixture to $75^{\circ}C$ to obtain clear liquor, then allowing the solution to cool slowly. Needle like crystals of monoammonium cyanurate monohydrate were obtained (Anal. Calcd for C$_3$H$_6$N$_4$O$_3$ · H$_2$O : C, 21.96; H, 4.91; N, 34.13. Found: C, 21.58; H, 4.87; N, 34.21).

Synthesis of Inorganic Reagents. To prepare the polysulfide solutions, a saturated solution of sodium sulfide nonahydrate (Na$_2$S·9H$_2$O) was mixed with reagent elemental sulfur and warmed to about $50^{\circ}C$. Dissolution was rapid and yielded a brownish orange solution of Na$_2$S$_n$, where $\underline{n}$ has a value of 2.32 (*16*).

Procedures. All experiments were conducted at ambient temperature (73 ± 2°F). Prior to each set of experiments, kinetic studies were done to determine the stirring times that would lead to maximum heavy metal removal. A sample of 10-34-0 was obtained from a commercial source. The WPA samples used in this study were prepared in TVA's Acid Research Facility (Kohler, J.J.; Tate, L.R., 194th National American Chemical Society Meeting, August 30-September 4, 1987, New Orleans, LA). In the first set of experiments, the organic reagent was added dropwise with stirring to a 50 g portion of either 10-34-0 or WPA over a 10 min period, and the resulting slurry was stirred for an additional 10 to 30 min. The solid and liquid phases were easily separated by filtration without the need for a flocculant, and the liquid phase was analyzed for heavy metals by inductively coupled plasma (ICP) spectroscopy. In the second set of experiments, the WI starch xanthate was added (amount added based on the assumed binding capacity and the moles of heavy metals present) to either 10-34-0 or WPA, and the resulting mixture was stirred for 2 h. The slurry was then filtered through a medium fritted-glass filter, and the liquid phase was analyzed for heavy metals using ICP.

In the third set of experiments, WPA containing 32% P$_2$O$_5$ and 1.3% SO$_3$ was treated with CdCO$_3$, HgCl$_2$, and PbCO$_3$ and diluted slightly to yield a solution containing, by calculation, 30% P$_2$O$_5$, 1.2% SO$_3$, and 300 ppm cadmium, mercury, and lead. After mixing overnight, the solution was clarified by centrifugation to remove any suspended solids. Solutions containing 18.1% Na$_2$S or 25.4% Na$_2$S$_{2.32}$ were then added to the WPA at rates of 1, 2, and 4 times the stoichiometric amount required for precipitation of the metals assuming the polysulfides to behave as single divalent anions. Mixtures were mechanically shaken for 30 minutes, centrifuged, decanted, and the liquor submitted for analysis along with a sample of the untreated clarified acid.

In the fourth set of experiments, ferrocyanide was examined as a precipitant for heavy metals in WPA. An organic flocculant (Calgon POL-E-Z 652) was added to improve the separation. WPA containing 33.2% P_2O_5 was treated with 20% $K_4Fe(CN)_6$ solution at rates equivalent to 20, 40, 75, and 150% of the soluble iron, assuming the iron to be present as iron(III) and the precipitate to have the composition $FeK[Fe(CN)_6]$. The flocculant was added as a 2% solution at a rate of 0.08 g solution per g potassium ferrocyanide. Poor separation occurred with the highest amounts of ferrocyanide addition and those precipitates were difficult to compact manually. These mixtures were therefore centrifuged to obtain solutions suitable for analysis. Only cadmium, nickel, and iron were analyzed in the treated samples from this experiment.

Finally, it should be noted here that highly acidic wastewaters (pH <1) will precipitate the acid form of TMT-15, trithiocyanuric acid. In fact, our preliminary tests confirmed that trithiocyanuric acid was indeed formed upon addition of TMT-15 to WPA (pH <1). Consequently, only the chemistry of TMT-15 with 10-34-0 will be discussed in a later section. The pH values of 10-34-0 are typically in the range of 6.2-6.9.

Use of Organic Reagents To Remove Heavy Metals From 10-34-0

The analytical results from the precipitation and adsorption tests conducted with TMT-15, Thio-Red II, 5% Na_2CS_3, and the WI starch xanthate on samples of 10-34-0 containing additional arsenic, cadmium, copper, lead, manganese, and mercury are summarized in Table II.

Trithiocyanuric Acid (TMT-15). Literature obtained from the manufacturer stated that TMT-15 would precipitate copper, cadmium, mercury, silver, lead, nickel, and tin from wastewater, after adjusting the pH value to within certain ranges (20). Trivalent metals, such as iron, chromium, and aluminum, will not precipitate upon addition of TMT-15.

The results obtained from initial tests where TMT-15 was added to 10-34-0 spiked only with additional cadmium were very impressive. The concentration of cadmium was reduced dramatically from 30 ppm to 0.70 ppm. Unfortunately, an expensive extraction procedure had to be utilized to concentrate the cadmium from the TMT-15 treated solutions to a level which could be analyzed using ICP spectroscopy. Therefore, the results described below were obtained from 10-34-0 which was spiked with cadmium and other heavy metals to reach a level adequate for direct ICP analysis.

The analytical results from subsequent precipitation tests conducted with TMT-15 (Table II) show that when one equivalent of TMT-15 was added for each equivalent of heavy metal present, copper (>99%), mercury (>95%), lead (79%), and cadmium (48%) precipitated from the 10-34-0. Relatively small amounts of arsenic (13%), zinc (<10%), chromium (<2%), and manganese (6%) precipitated upon addition of TMT-15. When a 50% excess of TMT-15 was used, only a small improvement in the precipitation of arsenic, cadmium, and lead was observed.

Table II. Analytical Results From Precipitation and Adsorption
Tests on 10-34-0 Spiked With Additional Heavy Metals[a]

Reagent	As	Cd	Cu	Hg	Mn	Pb	Cr	Zn
				Heavy Metals (ppm)				
Spiked 10-34-0	110	110	82	90	126	87	406	622
Thio-Red II								
0.50 equivalent	94	103	41	<4	126	75	406	602
1.00 equivalent	92	100	3	<4	126	68	404	581
1.50 equivalent	71	70	<1	<4	126	15	403	571
TMT-15								
0.75 equivalent	104	68	<1	<4	121	21	405	583
1.00 equivalent	96	57	<1	<4	118	18	403	568
1.25 equivalent	83	52	<1	<4	117	15	402	560
5% $Na_2 CS_3$								
0.50 equivalent	<1	<3	<1	<4	119	<10	388	574
1.00 equivalent	<1	<3	<1	<4	118	<10	387	284
1.50 equivalent	<1	<3	<1	<4	116	<10	390	105
WI Starch Xanthate								
Binding capacity A[b]	110	110	51	<4	124	66	404	620
Binding capacity B[c]	103	102	26	<4	123	40	400	617

[a]SOURCE: Reprinted with permission from ref. 19. Copyright 1990.
[b]1.1 mmole heavy metal/gram adsorbent.
[c]0.7 mmole heavy metal/gram adsorbent.

Since only 48% of the cadmium present in 10-34-0 precipitated in the above experiments, another test was performed to determine whether elevated levels of certain heavy metals were interfering with cadmium precipitation (e.g., 10-34-0 normally does not contain significant quantities of mercury, copper, nickel, and lead). In this test, the 10-34-0 was spiked only with additional cadmium, then sufficient TMT-15 was added to theoretically precipitate all the cadmium present. The solid phase was isolated and submitted for analysis. Once again, greater than 95% of the cadmium precipitated. Therefore, the high levels of other heavy metals which normally are not found in 10-34-0 were interfering with cadmium precipitation. In fact, Bhattacharyya and Ku (21) have found that nickel will interfere with cadmium precipitation when using inorganic sulfides. Consequently, if TMT-15 is used to precipitate cadmium, lead, and mercury from phosphate fluid fertilizers, a slight excess of TMT-15 will be required to precipitate each equivalent of heavy metal present.

Sodium Polythiocarbonate Solution (Thio-Red II). The analytical results from the precipitation tests conducted with Thio-Red II (Table II) show that when one equivalent of Thio-Red II was added for each equivalent of heavy metal present in the 10-34-0, only copper (>96%) and mercury (>95%) precipitated. Furthermore, only small amounts of lead (22%), arsenic (16%), and cadmium (9%) precipitated, while the levels of zinc (<7%), chromium (<1%), and manganese (<1%) in the 10-34-0 were essentially unaffected by addition of Thio-Red II. When a 50% excess of Thio-Red II was used, the amount of lead which precipitated increased to 83%, while the amounts of arsenic and cadmium which precipitated slightly increased to 34 and 36%, respectively.

Sodium Trithiocarbonate Solution (5% Na2CS3). Research by Elfine (*11,18*) showed that alkali metal trithiocarbonates, particularly sodium trithiocarbonate (Na2CS3), precipitated heavy metals from waste water as their insoluble sulfides. The claim that a low volume of sludge was generated by addition of the 5% Na2CS3 to waste water, as well as the claim that iron(III) and chromium(III) were precipitated, made this organic reagent a particularly attractive candidate for removing heavy metals from both 10-34-0 and WPA.

The analytical results from precipitation tests conducted with 5% Na2CS3 on 10-34-0 (Table II) were much more encouraging than the results from the precipitation tests with TMT-15 or Thio-Red II. The data given in Table II show that when one mole of 5% Na2CS3 was added for each mole of heavy metal present in the 10-34-0 (giving one CS_3^{2-} anion for each atom of heavy metal to be removed), arsenic (>99%), cadmium (>97%), copper (>98%), mercury (>95%), and lead (>94%) precipitated. On the other hand, the levels of chromium (<5%) and manganese (<6%) were virtually unaffected by addition of this reagent to the 10-34-0. When a 50% excess of 5% Na2CS3 solution was used, the amount of zinc which precipitated (83%) was quite high, but at an addition of one equivalent, only 54% of the zinc in the 10-34-0 was precipitated. Finally, even when a 50% excess of 5% Na2CS3 was used, the levels of chromium and manganese in the 10-34-0 were not appreciably affected.

The results of these precipitation tests with TMT-15, Thio-Red II, and 5% Na2CS3 can be understood by considering the K_{sp} values for related metal-ethyl xanthate complexes (*22*): copper and mercury have the smallest values, 10^{-20} and 10^{-38}, respectively; lead and cadmium are slightly higher, 10^{-17} and 10^{-14}, respectively; while the values for zinc and manganese are quite high, 10^{-9} and 10^2, respectively.

Cyanuric Acid. Organic cyanurates and isocyanurates have been prepared as pure materials and their use as metal precipitants has been reported in the literature (*23-25*). Diallylisocyanurate salts of cadmium, copper, and lead have been described and polyisocyanurates have been cited as precipitants for monovalent and divalent metal ions — including Cd, Hg and Pb — from waste streams (*26*). Initial tests with 10-34-0 (pH 6.8) indicated that ammonium cyanurate was soluble in the media; however, no measurement of the solubility was made and no precipitate was observed. Addition of ammonium cyanurate to a 10-30-0 (pH 6.0) grade phosphate fluid fertilizer containing 40 ppm cadmium indicated low solubility of the reactant in the media and resulted in no cadmium removal at stoichiometries ranging from 25 to 480%. Confirmatory tests using reagent cyanuric acid showed that it too was insoluble in 10-30-0 and resulted in no decrease in cadmium concentration.

WI Starch Xanthate. Research by Wing and others (*22, 27-29*) has shown that water-soluble (WS) starch xanthates, in combination with cationic polymers to form polyelectrolyte complexes, can effectively remove heavy metals from waste water. To eliminate the expensive cationic polymer and give a more economical method of heavy metal removal, further research by Wing and others (*12, 30-33*) showed that xanthation of a highly crosslinked starch yields a water-insoluble (WI) product that is effective in removing heavy metals from waste water without the need for a cationic polymer. In more recent work, Tare and Chaudhari (*34*) evaluated the effectiveness of the starch xanthate (WS and WI) process for removal of hexavalent chromium from synthetic waste waters.

The analytical results from adsorption tests conducted with the WI starch xanthate (Table II) show that at an addition of reagent corresponding to an assumed binding capacity of 1.1 mmole heavy metal per gram of adsorbent (*12*), only mercury (>95%), copper (38%), and lead (24%) were adsorbed. Similar to the results from the precipitation tests, levels of manganese and chromium in 10-34-0 were unaffected by this treatment. In contrast to the results from the precipitation test with the most effective liquid precipitant (5% Na_2CS_3), arsenic, zinc, and cadmium were not adsorbed from 10-34-0 by the WI starch xanthate. Furthermore, the amounts of copper and lead adsorbed were much less than the amounts of copper and lead precipitated in the previous tests. Even at an increased addition of xanthate corresponding to an assumed binding capacity of 0.7 mmole heavy metal per gram of adsorbent, the amounts of copper and lead adsorbed were still only 68 and 54%, respectively, while the adsorption of arsenic, cadmium, and zinc remained relatively low.

Use of Organic Reagents To Remove Heavy Metals From WPA

The analytical results from the precipitation and adsorption tests conducted with Thio-Red II, 5% Na_2CS_3, and the WI starch xanthate on WPA containing additional cadmium, copper, lead, manganese, and mercury are summarized in Table III. As mentioned previously, addition of TMT-15 to WPA results in the precipitation of trithiocyanuric acid; no heavy metals are coprecipitated.

Sodium Polythiocarbonate Solution (Thio-Red II). The analytical results from the precipitation tests conducted with Thio-Red II (Table III) show that when one equivalent of Thio-Red II was added for each equivalent of heavy metal present in the WPA, only copper (>98%), mercury (>97%), and lead (>75%) precipitated. The levels of cadmium (<8%), manganese (<2%), chromium (<1%), and zinc (<1%) were essentially unaffected by addition of Thio-Red II. The amount of these latter heavy metals which precipitated did not increase even when a 50% excess of Thio-Red II was used. Each mixture of WPA and Thio-Red II gave off a strong sulfide-type odor after stirring for 20 minutes, giving a qualitative indication that some decomposition of the reagent was occurring.

Sodium Trithiocarbonate Solution. The analytical results from the precipitation tests conducted with 5% Na_2CS_3 (Table III) show that when one mole of reagent was added for each mole of heavy metal present, copper (>97%), mercury (>97%), lead (>75%), and chromium (33%) precipitated from the WPA. Relatively small amounts of cadmium (14%), manganese (<7%), and zinc (<5%) precipitated. When a 50% excess of 5% Na_2CS_3 was used, slightly more cadmium (38%) and manganese (16%) precipitated from the WPA; however, precipitation of zinc was still negligible. Interestingly, when using identical amounts of 5% Na_2CS_3, the amount of chromium that precipitated from the WPA (33%) was significantly greater than the amount precipitated from 10-34-0 (<5%). Finally, a sulfide-type odor was not noticed during any of these tests, giving a qualitative indication that the 5% Na_2CS_3 reagent was not appreciably decomposed by the WPA.

WI Starch Xanthate. At an addition of WI starch xanthate to the WPA corresponding to an assumed binding capacity of 1.1 mmole heavy metals per gram of adsorbent, only mercury (>97%), copper (>97%), and lead (>75%) were effectively adsorbed (Table III). Adsorption of cadmium, chromium, manganese, and zinc was insignificant. No significant increase was noted in the amount of heavy metals adsorbed from the WPA even at an increased addition of xanthate corresponding to an assumed binding capacity of 0.7 mmole heavy metal per gram of adsorbent.

Table III. Analytical Results From Precipitation and Adsorption Tests on WPA Spiked With Additional Heavy Metals[a]

Reagent	Heavy Metals (ppm)						
	Cd	Cu	Hg	Mn	Pb	Cr	Zn
Spiked WPA	130	90	140	200	20	150	150
5% $Na_2 CS_3$							
0.50 equivalent	106	<1	<4	187	<5	97	143
1.00 equivalent	94	<1	<4	177	<5	100	142
1.50 equivalent	80	<1	<4	171	<5	98	131
Thio-Red II							
0.50 equivalent	130	6	<4	200	<5	150	150
1.00 equivalent	120	<1	<4	197	<5	149	149
1.50 equivalent	115	<1	<4	193	<5	146	147
WI Starch Xanthate							
Binding capacity A[b]	125	15	<4	189	<5	148	147
Binding capacity B[c]	121	<1	<4	186	<5	145	144

[a]SOURCE: Reprinted with permission from ref. 19. Copyright 1990.
[b]1.1 mmole heavy metal/gram adsorbent.
[c]0.7 mmole heavy metal/gram adsorbent.

Use of Inorganic Reagents To Remove Heavy Metals From WPA

Sulfide Precipitants. Both sodium sulfide and sodium polysulfide were tested for their ability to remove cadmium, mercury, and lead from WPA. The results are shown in Table IV. These results show that the removal of heavy metals was not greatly affected

Table IV. Sulfide Precipitation of Selected Heavy Metals From WPA

	Mole Ratio	Fraction Precipitated, wt %		
Precipitant	$Na_2Sn:P_2O_5$	Cd	Hg	Pb
Na_2S	0.0030	5	56	>50
Na_2S	0.0067	5	56	>50
Na_2S	0.0116	5	44	>50
$Na_2S_{2.32}$	0.0031	0	67	>50
$Na_2S_{2.32}$	0.0057	5	67	>50
$Na_2S_{2.32}$	0.0116	5	67	>50

by varying the amount of sulfide or polysulfide and that only minor removal of cadmium was obtained. This low degree of cadmium precipitation is, at first glance, surprising since Peters and Ku had shown that the concentration of cadmium sulfide is significantly lower than that of lead sulfide at low pH and both Maruyama and Berglund obtained low cadmium levels in WPA. However, Peters and Ku conducted their tests using waste water, which has very different chemical properties from WPA. Maruyama achieved high cadmium precipitation from WPA by using pressure to increase the sulfide concentration of the solution, while Berglund precipitated sulfate and partially neutralized the WPA prior to the sulfide treatment.

Although the polysulfide precipitated about 10 to 20% more mercury than the sulfide, about 30 ppm mercury still remained in solution after treatment. The effect on lead removal was not as clearly defined, even though over half the lead was removed in each test. Since the concentration of lead in the feed solution was already low at only 20 ppm and the lower detectable limit for the analytical method (i.e., ICP spectroscopy) was 10 ppm, the results demonstrated only that lead concentrations less than 10 ppm could be obtained by either treatment.

Ferrocyanide. Results from the ferrocyanide tests are shown in Table V. Iron is reduced in the approximate amounts determined by the design except in the last test where the residual iron is probably due to excess potassium ferrocyanide. The concentrations of both cadmium and nickel are reduced below the detectable limits of the analytical method even at the lowest stoichiometry tested. These results indicate that even lower amounts of ferrocyanide could produce substantial reduction in heavy metal concentrations. However, this method has two primary disadvantages: (1) the potential reaction of ferrocyanides with strong acids to form hydrocyanic acid (HCN) and (2) the

Table V. Ferrocyanide Precipitation of Iron
and Heavy Metals From WPA

Percent Stoichiometry	Solution Composition		
[Fe(CN)$_6$]:Soluble Fe	Fe, %	Cd, ppm	Ni, ppm
0	0.79	75	43
20	0.57	<10	<10
40	0.39	<10	<10
75	0.10	<10	<10
100	0.04	<10	<10

physical characteristics of the precipitate. The formation of HCN would likely be negligible when using WPA (30% P_2O_5) and operating at temperatures below 50°C as suggested by Wakefield et al. In fact, no evolution of gas was observed nor any odor of HCN detected in the current tests. Unfortunately, the extreme fineness and bulk of the precipitate make it difficult to recover the WPA quantitatively even with the use of flocculants. However, the lower stoichiometries suggested above would also result in substantially less precipitate, allow higher WPA recoveries, and require less recycle of ferrocyanide.

Use of Inorganic Reagents To Remove Heavy Metals From 10-34-0

Sulfide Precipitants. Sulfide precipitation of heavy metals from 10-34-0 prepared from Western U.S. superphosphoric acid was also investigated. Portions of the 10-34-0 were then treated with varying amounts of Na_2S and $Na_2S_{2.32}$ solutions. The results of these tests are shown in Table VI.

Varying degrees of precipitation of arsenic, cadmium, mercury, and lead were obtained. At the highest levels of removal the concentrations of these four metals were all below 10 ppm. While chromium was unaffected under all conditions, iron was slightly affected, although this was insignificant in terms of its total concentration. Based on the fraction of metal precipitated at each sulfide level, the order of precipitation was found to be Hg>As>Pb>Cd>Fe>Cr.

The data given in Table VI for removal of arsenic, cadmium, and lead using polysulfide were almost twice that obtained with sulfide at the lower treatment levels. This higher response to polysulfide may be due to the lower solubility of the polysulfide salts, coordination compounds having lower solubilities, or the polysulfides may simply furnish higher effective sulfide concentration. The stoichiometry of the precipitated material was not determined in these tests.

Summary

Organic Reagents. Arsenic, cadmium, copper, mercury, lead, and to a lesser extent, zinc, can be precipitated from 10-34-0 by adding the 5% Na_2CS_3 reagent. Manganese, iron, and chromium do not precipitate upon the addition of this reagent. The 5% Na_2CS_3 reagent was identified as the most effective organic reagent tested during the course of

Table VI. Sulfide Precipitation of Heavy Metals From 10-34-0

	Mole Ratio	Fraction Removed, wt %					
Precipitant	$Na_2S_n:P_2O_5$	As	Cd	Cr	Hg	Pb	Fe
Na_2S	0.0085	5	0	0	82	0	0
Na_2S	0.0142	10	0	0	89	2	0
Na_2S	0.0285	45	51	0	>95	67	1
Na_2S	0.0563	>97	>96	0	>95	>96	2
$Na_2S_{2.32}$	0.0074	21	0	0	>95	4	0
$Na_2S_{2.32}$	0.0154	31	20	0	>95	53	2
$Na_2S_{2.32}$	0.0297	90	>96	0	>95	>96	0
$Na_2S_{2.32}$	0.0549	>95	>96	0	>95	>96	3

this project for removing heavy metals from 10-34-0. This technique should be directly applicable to other grades of phosphate fluid fertilizers as well.

Mercury, copper, lead, and cadmium can be precipitated from 10-34-0 by adding TMT-15. The TMT-15 and Thio-Red II precipitated at least 94% of the copper and mercury present in the 10-34-0; manganese and chromium were not precipitated. The WI starch xanthate adsorbed mercury, copper, and lead from 10-34-0, while the adsorption of arsenic, cadmium, chromium, manganese, and zinc was negligible.

Mercury, copper, lead, and to a lesser extent, chromium and cadmium, can be precipitated from WPA by adding the 5% Na_2CS_3 reagent. This reagent does not precipitate manganese or zinc from WPA. Unfortunately, a strong sulfide-type odor was evolved from the mixture of WPA and Na_2CS_3 giving a qualitative indication that the reagent was decomposing. Thio-Red II also precipitated mercury, copper, and lead from the WPA; but this reagent did not precipitate cadmium, manganese, chromium, or zinc. TMT-15 decomposed in WPA to give trithiocyanuric acid. The WI starch xanthate adsorbed mercury, copper, and lead from WPA.

Inorganic Reagents. The removal of arsenic, cadmium, and lead from 10-34-0 using polysulfide was almost twice that obtained with sulfide. The polysulfides appear to be promising for removal of several heavy metals from phosphate fluid fertilizers but are not satisfactory for use with WPA. However, ferrocyanide was shown to be highly effective for use with WPA. Ferrocyanide precipitated iron, cadmium, and nickel from WPA to below the detectable limits of the analytical method. The possible formation of hydrocyanic acid and the extreme fineness and bulk of the metal-ferrocyanide precipitate are disadvantages of this method.

Literature Cited

1. Johnson, F. J. *Solutions* **1989**, *33(2)*, 46.
2. COM(87) 165 Final, Commission of the European Communities, Brussels (21 April 1987).
3. *European Chemical News*, 25 (27 April 1987).

4. *Phosphorus & Potassium 149,* 40 (May-June 1987).

5. *Phosphorus & Potassium 162,* 23 (July-August 1989).

6. *Phosphorus & Potassium 130,* 35 (March-April 1984).

7. *Phosphorus & Potassium 137,* 26 (May-June 1985).

8. *Phosphorus & Potassium 147,* 30 (January-February 1987).

9. Bouffard, L.E.; Morris, S.B. AIChE Joint Meeting, Clearwater, FL (May 1988).

10. Wing, R. E.; Rayford, W.E. *Plat. and Surf. Finish* **1982**, *69,* 67.

11. Elfine, G. S. In U.S. Patent 4,678,584 (July 7, 1987).

12. Wing, R. E.; Doane, W. M.; Russell, C.R. *J. Appl. Polymer Sci.* **1975**, *19,* 847.

13. Peters, R. W.; Ku, Y. In *Separation of Heavy Metals and Other Trace Contaminants;* Peters, R. W., Ed.; AIChE Symposium Series; American Institute of Chemical Engineers: New York, NY, 1985, Vol. 81; pp 9-27.

14. Maruyama, K. In Japanese Patent 50/75115 (1975).

15. Berglund, H. A. In PCT Int. Appl. WO 80/2418 (1980).

16. Pickering, T. L.; Tobolsky, A. V. In *Inorganic and Organic Polysulfides;* Senning, A., Ed.; Sulfur in Organic and Inorganic Chemistry; Marcel Dekker, Inc.: New York, NY, 1972, Vol. 3; pp 19-38.

17. Harre, E. A.; Hargett, N. L. *Fertilizer Focus* **1987**, *4,* 46.

18. Elfine, G. S. In U.S. Patent 4,612,125 (September 16, 1986).

19. Norwood, III, V. M.; Kohler, J. J. *Fertilizer Research* **1990**, *26,* 113.

20. Reimann, D. O. *Waste Manag. and Res.* **1987**, *5,* 147.

21. Bhattacharyya, D.; Ku, Y. U.S. Environmental Protection Agency, EPA-600/S2-84-023 (1984).

22. Wing, R. E.; Swanson, C. L.; Doane, W. M.; Russell, C. R. *J. Water Pol. Contr. Fed.* **1974**, *46,* 2043.

23. Smolin, E. M.; Rapoport, L. In *The Chemistry of Heterocyclic Compounds;* Weissberger, A., Ed.; s-Triazines and Derivatives; Interscience: New York, NY, 1959, Vol. 13; pp 17-48.

24. Taylor, R. M. *Z. Anorg. Allg. Chem.* **1972**, *390,* 85.

25. Clyde, D. D. *J. Chem. Ed.* **1988**, *65,* 911.

26. Agrabright, P. A.; Echelberger, L. M.; Phillips, B. L. In U.S. Patent 3,893,916 (July 8, 1975).

27. Wing, R. E. In U.S. Patent 3,979,286 (September 7, 1976).

28. Swanson, C. L.; Wing, R.E.; Doane, W.M.; Russell, C.R. *Envir. Sci. Technol.* **1973**, *7,* 614.

29. Marani, D.; Mezzana, M.; Passino, R.; Tiravanti, G. *Envir. Technol. Lett.* **1980**, *1,* 141.

30. Wing, R. E.; Doane, W. M. *Plat. and Surf. Finish* **1981**, *68,* 50.

31. Wing, R. E.; Rayford, W. E.; Doane, W. M.; Russell, C. R. *J. Appl. Poly. Sci.* **1978**, *22,* 1405.

32. Rayford, W. E.; Wing, R. E.; Doane, W. M. *J. Appl. Poly. Sci.* **1979**, *24,* 105.

33. Campanelli, L.; Cardarelli, E.; Ferri, T.; Petronio, B. M. *Wat. Res.* **1986**, *20,* 63.

34. Tare, V.; Chaudhari, S. *Wat. Res.* **1987**, *21,* 1109.

RECEIVED February 20, 1992

Chapter 12

Kinetics of Removal of Heavy Metals by a Chelating Ion-Exchange Resin

Regeneration of the Resin by NH$_4$OH Solution

K. C. Kwon[1], Helen Jermyn[2], and Howard Mayfield[2]

[1]Chemical Engineering Department, School of Engineering and Architecture, Tuskegee University, Tuskegee, AL 36088
[2]Environics Division, Air Force Civil Engineering Support Agency, HQ AFCESA/RAVC, Tyndall Air Force Base, FL 32403—6001

Experiments were performed for the removal of Cr, Fe^{+2}, Cu^{+2}, and Ni^{+2} with Dowex XFS 4195.02 ion exchange resin at 25°C. A mathematical model for the removal of both Cr and Cu^{+2} from aqueous solutions in the presence of the resin was developed with the assumption that intraparticle diffusion is a controlling step for the adsorption of the heavy metal ions on the resin. The intraparticle diffusivities of Cr and Cu^{+2} through the resin were obtained to be 0.5 and 0.03 - 0.05 cm^2/min, respectively. A series of experiments on the regeneration of Cu^{+2} saturated resin were conducted at 25°C, using 0.5 - 2 N NH$_4$OH aqueous solution. A mathematical model on the regeneration of the exhausted resin was developed.

Ion exchange resins have been utilized in the treatment of contaminated water, and in hydrometallurgical processes for the production of copper, nickel, uranium, and cobalt. More recently, chelating ion exchange resins have been used for the purification of plating bath solutions. During the plating operation, plating baths and rinsewaters are contaminated with excess metals such as chromium, copper, iron, aluminum, lead, zinc, and others. These contaminant metals reduce the efficiency of the plating operations. In order to maintain efficiency, the plating solutions must be replaced periodically. Disposing of the spent plating solutions, with the contaminating metals, is complicated by federal environmental regulations restricting hazardous waste

0097—6156/92/0509—0161$06.00/0

discharges. Plating bath solutions must be disposed of as hazardous wastes, at a cost which is approximately $1,000 per ton.

The current practice of periodically replacing spent plating bath solutions is doubly expensive, on one hand due to the expense of fresh plating bath solution, and on the other due to the cost of disposing of the waste. A desirable alternative approach is to regenerate the plating bath solution by selectively sorbing the contaminants out of the bath. Various ion exchange resins, such as Amberlite IRC-718, Duolite ES-467, Dowex XFS-43084, Dowex XFS-4195, Unicellex UR-10 and Sumichelate CR-2 have been used experimentally to restore efficiencies of metal-contaminated plating bath solutions (1).

Chelating resins have proved effective in the removal of metals from plating bath solutions provided that the resins are fresh or effectively regenerated. Care may be required in the regeneration procedure to ensure that plating bath solutions treated with the regenerated resin are not contaminated by metals remaining in the resin after regeneration (2).

Dowex XFS 4195 is a chelating ion exchange resin consisting of macroporous polystyrene/divinylbenzene copolymer onto which weakly basic chelating picolylamine derivatives have been attached (3). This resin has been found to be effective in the removal of contaminated metals from metal plating bath solutions when the fresh resin was used. Problems have occurred from the contamination of plating baths with metals from regenerated Dowex 4195. When the resins are saturated with contaminant ions, the plating bath itself may leach additional contaminant ions from the resin. Thus, treating the plating bath with used resin may increase the concentration of the contaminants and defeat the purpose of the resin treatment. There is a need for further studies on resin regeneration in order to develop effective techniques for re-using the resin(2).

The successful use of chelating ion exchange resins to treat metal plating baths depends upon the selective sorption properties of the resins for the ions present in the plating baths. The selectivity of the sorption can depend upon a number of factors, including the temperature, pH, and upon the type and concentration of other anions and cations present in the solution. Various competing ion effects have been noted for metals such as copper, nickel, cobalt, iron, zinc and others (4 - 8).

The chelating ion exchange resins are usually regenerated for re-use by eluting or extracting the sorbed metal from the resin. Both acidic solutions, such as HCl and H_2SO_4, and basic solutions, such as aqueous ammonia, have been used to regenerate loaded ion exchange resins. Sulfuric acid solutions of various strengths have been reported to remove most metals except copper from XFS 4195 resin. The removal of copper from XFS 4195 has best been accomplished with aqueous ammonia (4, 7 - 9).

In this present research, Dowex XFS-4195.02 was selected as a representative chelating ion exchange resin to be used for the removal of heavy metals. Copper, chromium, nickel, and iron were chosen as representatives of contaminant heavy metals in plating bath solutions. Aqueous ammonia was used as a regenerant for copper-saturated XFS 4195 resin. Special emphasis was given to identifying the kinetics of removal of Cu^{+2} from aqueous solutions by fresh resin and the kinetics of regeneration of the copper-saturated resin. The principal objectives of this current work are (1) to identify optimum conditions of regeneration for the used Dowex XFS 4195.02 ion exchange resin, (2) to study the kinetics on the regeneration of the Dowex XFS 4195.02 ion exchange resin loaded with heavy metals, and (3) to study the kinetics of the metal removal from aqueous solutions in the presence of Dowex XFS 4195.02 resin. Aqueous ammonia solutions of various strengths, and other regeneration solutions and conditions were studied in search of the optimum conditions. The kinetics of the metal removal from aqueous media by fresh resin were studied under various treatment conditions.

Mathematical models for the removal of heavy metals from aqueous solutions and the regeneration of used Dowex XFS 4195.02 resin were developed under several assumptions. Intraparticle diffusion is assumed to be the controlling step for the removal of heavy metals from aqueous solutions (10). The reaction rate is assumed to be a controlling step for the regeneration of the loaded resin in conjunction with a mass balance of heavy metal ions in the batch reactor (11,12).

The quasi-steady state assumption was made for the development of mathematical models. The particle size of the resin is assumed to be not changeable during the adsorption of heavy metals on the resin as well as during the regeneration of the loaded resin. The assumptions of isothermal behavior and absence of swelling changes are less justified, since the ionic reactions may release considerable heat and may make the ion exchanger swell or shrink appreciably during ion exchange processes (13).

A mathematical model on the removal of heavy metals from aqueous solutions in the presence of Dowex XFS 4195.02 resin was developed, as shown in Equation 1.

$$-\frac{4\pi D_e A R t}{V} = \ln\left[\frac{Y^2 + AY + A^2}{1 + A + A^2}\right]^{-0.5} + \ln(1 - X_A)^{-A} + \ln\left[\frac{Y - A}{1 - A}\right] + \sqrt{3}\left[\tan^{-1}\left[\frac{2Y + A}{\sqrt{3}A}\right] - \tan^{-1}\left[\frac{2 + A}{\sqrt{3}A}\right]\right] \quad (1)$$

The term A is defined in Equation 2.

$$A = \left[1 - \frac{C_{Ao} V b}{W}\right]^{1/3} \quad (2)$$

Y is given by Equation 3.

$$Y^3 = 1 - (1 - A^3) X_A \tag{3}$$

Where V is the volume of aqueous metal solution, W is the amount of resin, L_B is the density of the dry ion exchange resin, R is the radius of the spherical resin, B is the adsorption capacity of resin, in g-mole-metal-ion/g-resin, t is the adsorption time, D_e is the effective intraparticle diffusion of heavy ion metals through resins, X_A is the removal fraction of heavy metal ions by ion exchange resins, and C_{Ao} is the initial concentration of metal ions.

A mathematical model for the regeneration of Cu^{+2}-saturated Dowex XFS 4195.02 ion exchange resin was developed, assuming that reaction rate is a controlling step for the regeneration of the Cu^{+2}-saturated resin in the presence of 1 - 2 N aqueous ammonia solution, as shown in Equation 4.

$$-\ln \left[\frac{X_e - X}{X_e - X_o} \right] = kt \tag{4}$$

Where X is the regeneration fraction, t is the regeneration time, in minutes, X_o is the instantaneous regeneration fraction, X_e is the equilibrium regeneration fraction, and k is the regeneration rate constant, in min^{-1}.

A mathematical model for the regeneration of Cu^2-saturated Dowex XFS 4195.02 ion exchange resin was developed, assuming that intraparticle diffusion of Cu^{+2} through the resin is a controlling step in the presence of 0.5 N aqueous ammonia as a regenerant, as shown in Equation 5.

$$1 - Y^{-1} - lnY = \frac{4\pi D_e R}{3V} t \tag{5}$$

Where Y for Equation 5 is given by Equation 6.

$$Y^3 = 1 - \frac{VC_A}{WB} \tag{6}$$

Where t is the regeneration time, in minutes, V is the volume of aqueous regeneration solution, W is the amount of resin, R is the radius of a spherical resin bead, B is the adsorption capacity of the resin, in g-mole-metal-ion/g-resin, D_e is the effective intraparticle diffusivity of the heavy metal ion through the resin, and C_A is the concentration of metal ion in the aqueous regeneration solution.

Experimental

Dowex XFS-5195.02 was chosen as a representative of the chelating ion exchange resins. Copper, chromium, nickel, and iron were chosen as representative contaminants for laboratory bench-scale studies. Experimental runs on both the removal of the contaminant metals and the regeneration of the spent resin were performed in appropriately sized 100 - 250 mL glass beakers. For metal removal studies, each container was charged with 20 - 120 mL quantities of aqueous solutions containing 40 - 120 ppm of the metals studied, plus 0.1 - 0.6 g loadings of Dowex XFS 4195.02 resin. All solutions were maintained nominally at 25°C by room temperature. A Teflon-coated magnetic stirrer was added to each beaker to ensure homogeneity of the solutions. Samples of the solutions were drawn off periodically and analyzed by flame atomic absorption spectrometry, using an air-acetylene flame.

Regeneration studies were conducted by charging glass beakers with 0.2 - 0.8 g quantities of used Dowex XFS 4195.02 resin and 80 mL of aqueous ammonia at concentrations ranging from 0.5 - 2.0 N. A magnetic stirrer ensured homogeneous ammonia concentrations throughout the solutions, and the solutions were maintained nominally at 25°C by room temperature. Samples of the solutions were drawn off periodically and analyzed for metals using flame atomic absorption spectrometry with an air-acetylene flame.

Results and Discussion

Removal Studies

No significant removal of iron was observed when 82 g of aqueous solution, containing 86 ppm Fe^{+2}, was treated with 0.1 g of the Dowex XFS 4195.02 resin.

Experiments were performed on the removal of Cu^{+2} from aqueous solution at pH 1.5 and 25°C with the Dowex XFS 4195.02. These data allowed the development of a mathematical equation describing the isotherm equilibrium adsorption of Cu^{+2}, using the Freundlich equation. These results are shown in Figure 1.

In another series of experiments, varying amounts of resin (0.1 - 0.3 g) were equilibrated with 80 g aliquots of aqueous solution containing 80 ppm Cu^{+2} in order to identify effects of the resin loading on the rate of Cu^{+2} removal. These results are shown in Figure 2. Removal fractions of Cu^{+2} increase with increasing amounts of resin. These data were fitted to the mathematical model on the removal of Cu^{+2} from aqueous solution in order to obtain the intraparticle diffusivities of Cu^{+2} through the resin during removal from aqueous media. The solid lines in Figure 2 are predicted fractional removal of Cu^{+2} from Equation 1 and the calculated intraparticle diffusivity values. The intraparticle diffusivities appear to increase with increasing amounts of resin.

The effects of varying initial Cu^{+2} concentration on the removal of Cu^{+2} were examined by exposing 0.2 g of resin with 80 g quantities of solutions containing 40 - 120 ppm Cu^{+2}. The experimental data were

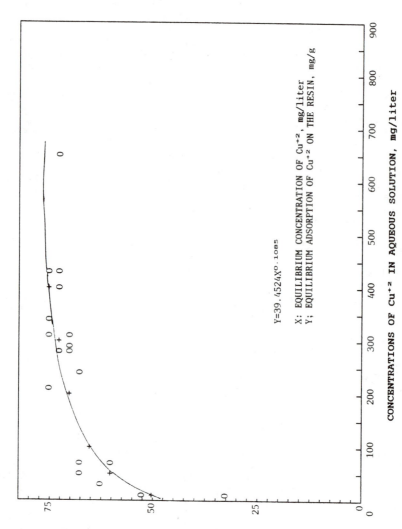

Figure 1. Isotherm equilibrium adsorption of Cu^{+2} on Dowex XFS 4195.02 ion exchange resin at 25°C.

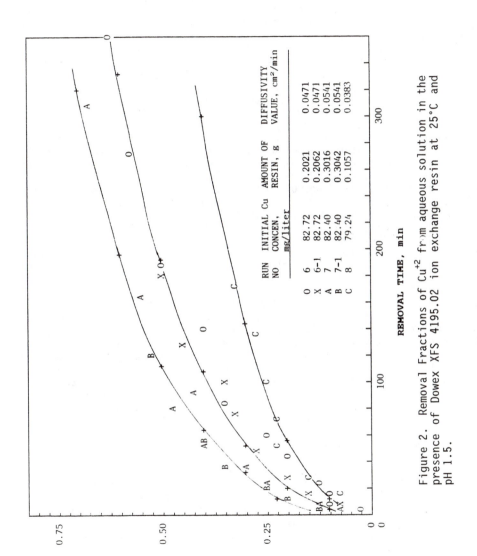

Figure 2. Removal Fractions of Cu^{+2} from aqueous solution in the presence of Dowex XFS 4195.02 ion exchange resin at 25°C and pH 1.5.

RUN NO	INITIAL Cu CONCEN, mg/liter	AMOUNT OF RESIN, g	DIFFUSIVITY VALUE, cm²/min
O 6	82.72	0.2021	0.0471
X 6-1	82.72	0.2062	0.0471
A 7	82.40	0.3016	0.0541
B 7-1	82.40	0.3042	0.0541
C 8	79.24	0.1057	0.0383

REMOVAL TIME, min

REMOVAL FRACTIONS OF Cu⁺²

applied to the removal model in order to identify the right-side values of the removal model, as is shown in Figure 3. The solid line in Figure 3 represents a least-squares fit between the right-side values of the removal model and the corresponding time data. The intraparticle diffusivities of Cu^{+2} through the resin were obtained from the slope of the solid line in Figure 3. Intraparticle diffusivities appear to increase with the initial concentration of Cu^{+2}, while the amounts of resin and solution are held constant.

The effects of the initial Cu^{+2} concentration upon the removal fraction are shown in Figure 4. In Figure 4, the symbols are the data, and the solid lines are the best-fit prediction from the removal model. Removal fractions increase with decreasing initial Cu^{+2} concentration. These observations indicate that a controlling step for the removal of Cu^{+2} from aqueous solution is diffusion of Cu^{+2} through the resin.

The effects of varying the relative amounts of solution and resin on the removal of Cu^{+2} were examined by treating varying quantities of 80 ppm Cu^{+2} solution with 0.2 g fixed quantities of the Dowex XFS 4195.02 resin at 25°C. The removal-fraction versus removal-duration data were applied to the removal model in order to identify effects of varying the proportions of solution and resin on the removal of Cu^{+2}. As is shown in Figure 5, the intraparticle diffusivities of Cu^{+2} through the resin increase with the amounts of solution. This may indicate that increased mass ratios of solution to resin result in increased Cu^{+2} removal.

In order to evaluate the effects of other transition metal ions on the removal of Cu^{+2}, solutions containing both Cu^{+2} and Ni^{+2} were treated with the Dowex XFS 4195.02 resin. Data from these experiments are shown graphically in Figure 6. The removal of Cu^{+2} from aqueous solutions is slightly retarded by the presence of Ni^{+2}.

In order to evaluate the utility of the removal model for additional metals, additional removal studies were conducted using Chromium and Cobalt in place of copper. Chromium removal was studied using solutions made by dissolving CrO_3 in dilute H_2SO_4 to produce solutions with chromium concentrations of 80 ppm. The oxidation state of the chromium in the resulting solutions was not determined. Intraparticle diffusivities of Cr, Co^{+2}, and Cu^{+2} were calculated with the removal model, and are compared in Figure 7. The removal model for Dowex XFS 4195.02 appears to be generally applicable to most metals.

Regeneration Studies

The regeneration of copper-saturated Dowex resin with aqueous ammonia was studied, using ammonia solutions with NH_4OH concentrations from 0.5 - 2.0 N. The relative amounts of resin and regenerating solution were also varied. These results are summarized in Figure 8. Equilibrium concentrations of Cu^{+2} in aqueous ammonia solution decrease with increasing amounts of copper-saturated resin, and appear to be independent of the ammonia concentration.

A series of experiments were carried out on the regeneration of copper-saturated Dowex XFS 4195.02 resin. The trials were conducted

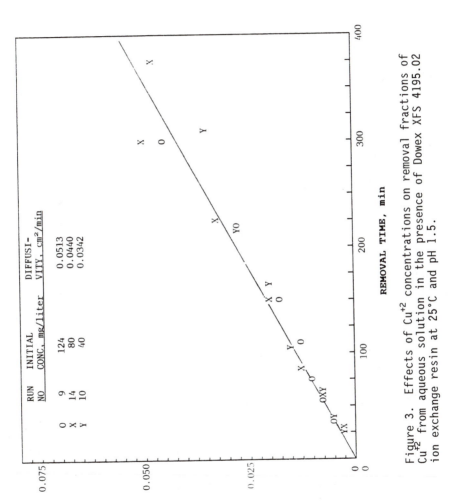

Figure 3. Effects of Cu^{+2} concentrations on removal fractions of Cu^{+2} from aqueous solution in the presence of Dowex XFS 4195.02 ion exchange resin at 25°C and pH 1.5.

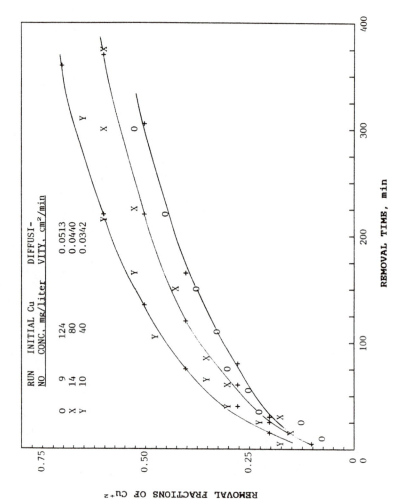

Figure 4. Effects of Cu^{+2} concentrations on removal fractions of Cu^{+2} from aqueous solution in the presence of Dowex XFS 4195.02 ion exchange resin at 25°C and pH 1.5.

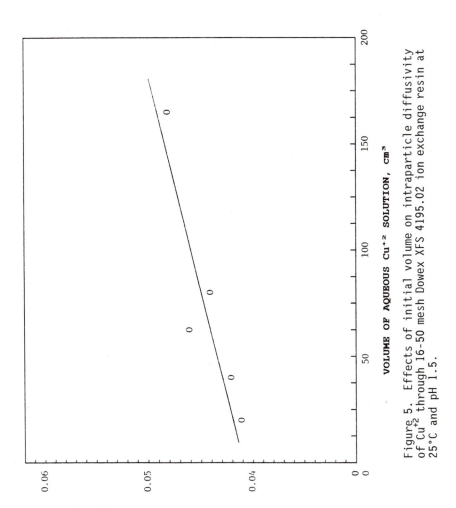

Figure 5. Effects of initial volume on intraparticle diffusivity of Cu^{+2} through 16–50 mesh Dowex XFS 4195.02 ion exchange resin at 25°C and pH 1.5.

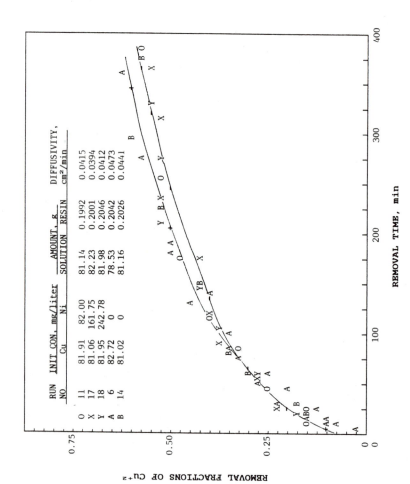

Figure 6. Effects of Ni^{+2} on the removal of Cu^{+2} from aqueous solution in the presence of Dowex XFS 4195.02 ion exchange resin at 25°C and pH 1.5.

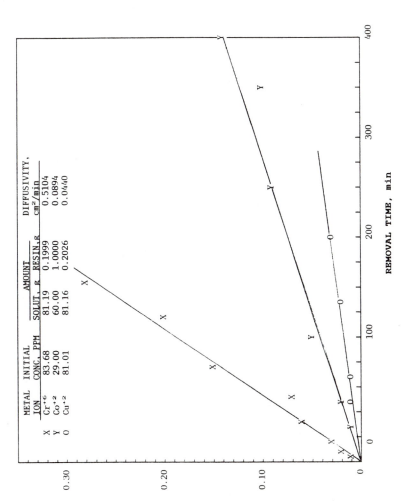

Figure 7. Comparison of intraparticle diffusivities of heavy metal ions through Dowex XFS 4195.02 ion exchange resin.

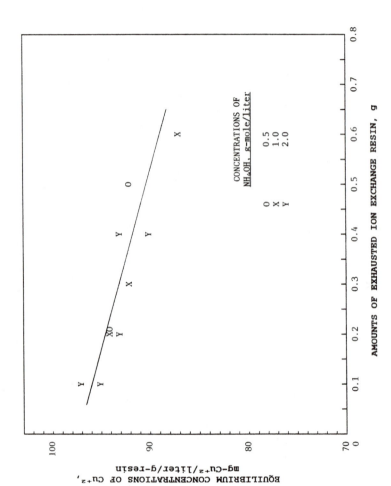

Figure 8. Equilibrium concentrations of Cu^{+2} in aqueous ammonia solution during regeneration of spent Dowex XFS 4195.02 ion exchange resin.

at room temperature (25°C) using aqueous ammonia solutions with NH_4OH concentrations ranging from 0.5 - 2.0 N as the regenerating reagent. The regeneration of the resin may be controlled by reaction rates or by intraparticle diffusion. Cases where the regeneration is controlled by reaction rates are modelled by Equation 4, and cases controlled by intraparticle diffusion are modelled by Equation 5. Instantaneous regeneration fractions and regeneration rate constants were obtained in reaction controlled cases from the intercepts and slopes of the linear plots obtained by applying Equation 4. Intraparticle diffusivities of Cu^{+2} through the resin were obtained for diffusion controlled cases by applying Equation 5.

The regeneration rate constants of Cu^{+2}-saturated resin in the presence of aqueous ammonia solutions are independent of resin amounts and dependent upon NH_4OH concentration. Instantaneous regeneration fractions of the resin are independent of resin amounts and NH_4OH concentrations in reaction controlled cases, shown in Figure 9. The reaction rate controlled model is applicable to the regeneration of copper-saturated Dowex XFS 4195.02 resin in solutions with NH_4OH concentrations ranging from 1 - 2 N.

The controlling step for the regeneration of copper-saturated Dowex 4195.02 resin in 0.5 N NH_4OH is dependent upon the resin amounts relative to the amount of regenerating solution. The regeneration data follow the reaction controlled model (Equation 4) in the case of 0.2 g of resin regenerated with 80 g of ammonia solution. The regeneration data follow the diffusion controlled model (Equation 5) in the cases where 0.3 g and 0.6 g of resin are regenerated with 80 g of ammonia solution. This is shown in Figure 10, where the plot deviates from a straight line in the 0.2 g case. This suggests that the controlling step for the removal of Cu^{+2} from copper-saturated resin is dependent upon the mass ratio of spent resin to ammonia solution.

Conclusions

The following conclusions were drawn on the basis of experimental data for the removal of heavy metals from aqueous solutions in the presence of Dowex 4195.02 ion exchange resin, and the regeneration of spent Dowex XFS 4195.02 in the presence of aqueous ammonia as a regenerant. (1) The removal of Cu^{+2} from aqueous solutions is controlled by intraparticle diffusion of Cu^{+2} through the resin. Removal rates of Cu^{+2} from aqueous solutions increase with initial concentrations of Cu^{+2}, and are independent of solution quantities. (2) Removal rates of Cu^{+2} from aqueous solutions decrease slightly in the presence of Ni^{+2}, but are otherwise independent of Ni^{+2} concentration. (3) Regeneration rates of spent resins are dependent on NH_4OH concentrations. The reaction rates are dependent on the mass ratios of ammonia solution to resin at low NH_4OH concentrations, but are independent of the mass ratios at high NH_4OH concentrations. (4) Instantaneous regeneration fractions of spent resins are independent of mass ratios and NH_4OH concentrations. Regeneration rate constants are independent of mass ratios and dependent on NH_4OH concentrations.

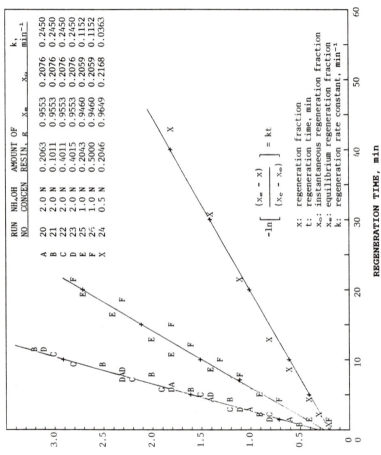

Figure 9. Regeneration of Cu^{+2}-saturated Dowex XFS 4195.02 ion exchange resin in the presence of aqueous ammonia at 25°C.

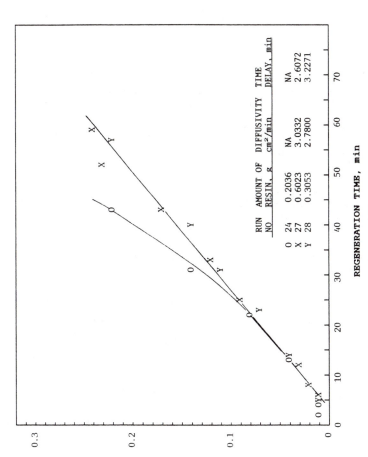

Figure 10. Effects of amounts of Cu^{+2}-saturated Dowex XFS-4195.02 ion exchange resin, in fixed quantities of regenerating solution, on the regeneration mechanism in the presence of aqueous ammonia solution at 25°C.

Acknowledgements

This research was supported by the U. S. Air Force Office of Scientific
Research and the U. S. Air Force Civil Engineering Support Agency.

Disclaimer

Mention of commercial equipment, instruments, or materials in this
paper does not imply recommendation or endorsement by the Air Force,
nor does it imply that the materials or equipment are necessarily the
best available for the purpose.

Literature Cited

1.   Gold, H.; Czupryna, G.; Robinson, J. M.; Levy, R. D. *Purifying
Air Force Plating Baths by Chelate Ion Exchange*, Final Report to
Headquarters, Air Force Engineering and Service Center, Tyndall AFB,
FL, Foster-Miller, Inc., July 1985.
2.   Folson, D. W.; Langham, A. *Selective Ion Exchange Pilot Plant
Demonstration*, Final Report to Headquarters, Air Force Engineering and
Services Center, Tyndall AFB FL Battelle Columbus Division, January
1989.
3.   Rosato, L.; Harris, G. B.; and Stanley, *Hydrometallurgy*, 1984,
vol 13, pp 13-44.
4.   Grinstead, R. R.; Nasutavicus, W. A.; Wheaton, R. M. In
Extractive Metallurgy of Copper; Yannopoulos, J. C.; Agarwal, J. C.,
Eds; The Metallurgical Society of AIME: New York, NY, 1976, Vol. 2; pp
1009-1024.
5.   Grinstead, R. R. New Developments in the Chemistry of XFS 4195
and XFS 42084 Chelating Ion Exchange Resins, In *Ion Exchange
Technology*; Naden, D.; Streat, M., Eds; Society of Chemical Industry:
London, 1984; p 509.
6.   Price, S. G.; Heditch, D. J.; Cegb; Streat, M. Diffusion or
Chemical Kinetic Control in a Chelating Ion Exchange Resin System, In
Ion Exchange for Industry; Streat, M., Ed; SCI, Ellis Horwood.
7.   Jeffers, T. H.; Harvey, M. R., *Cobalt Recovery from Copper Leach
Solutions*, U. S. Bureau of Mines Report of Investigations 8927, 1985.
8.   Grinstead, R. R. *Hydrometallurgy*, 1984, Vol. 12, pp 387-400.
9.   Jeffers, T. H. *J. Met.*, 1985, Vol. 37, pp 47-50.
10.   Yoshida, H.; Kataoka, T.; Ikeda, S. *Canadian J. Chem. Eng.*, 1985,
Vol. 63, pp 430-435.
11.   Levenspiel, O. *Chemical Reaction Engineering*, 2nd ed., John Wiley
& Sons: New York, NY, 1972.
12.   Himmelblau, D. A. *Basic Principals and Calculations in Chemical
Engineering*, 4th ed., Prentice-Hall: Englewood Cliffs, NJ, 1982.
13.   Helfferich, F. *J. Phys. Chem.*, 1965, Vol. 69, pp 1178.

RECEIVED February 20, 1992

FUTURE DIRECTIONS

Chapter 13

Ligand-Modified Micellar-Enhanced Ultrafiltration for Metal Ion Separations

Use of N-Alkyltriamines

Donald L. Simmons[1,2], Annette L. Schovanec[1,2], John F. Scamehorn[2,3], Sherril D. Christian[1,2], and Richard W. Taylor[1,2,4]

[1]Department of Chemistry, [2]Institute for Applied Surfactant Research, and [3]School of Chemical Engineering and Materials Science, University of Oklahoma, Norman, OK 73019

Ligand-modified micellar-enhanced ultrafiltration (LM-MEUF) is a membrane-based separation technique that utilizes an amphiphilic ligand, solubilized in micelles, to selectively complex a target metal ion in a mixture of metal cations. The ligands tested were N-(n-dodecyl)-diethylenetriamine (DDIEN), N-(n-dodecyl)-di-2-picolylamine (DDPA), and N-(n-hexadecyl)-di-2-picolylamine (HDPA). These compounds were tested with the surfactants cetylpyridinium chloride (CPC), cetyltrimethylammonium bromide (CTAB), N,N-dimethyl-dodecylamine-N-oxide (DDAO), polyoxyethylene nonyl phenyl ether $(NP(EO)_{10})$, and sodium dodecyl sulfate (SDS). All three ligands solubilized readily into each type of surfactant. Solubilization parameters showed a strong dependence on the charges of the surfactant and ligand, and the length of the ligand alkyl chain. With HDPA, dialysis measurements with Cu^{2+}/Ca^{2+} mixtures gave rejection values for Cu^{2+} as high as 99.8% (CPC), 99.7% (CTAB), 96.7% (DDAO), and 98.9% $(NP(EO)_{10})$, with no rejection of Ca^{2+}. Addition of salt (NaCl) increased the retention of the bound Cu^{2+}, but, when used with cationic surfactants, diminished the expulsion of Ca^{2+} into the permeate.

A number of membrane-based separation techniques have been developed wherein a surfactant is added to a solution containing some target species (cation, anion, molecule). The solution is treated in an ultrafiltration unit with membrane pore sizes sufficiently small to reject the micellar aggregates. Organic species solubilized in the micelle or ionic species associated with the micelle remain in the retentate (1-15). Thus, micellar-enhanced ultrafiltration (MEUF) provides an effective way to reduce the level of contaminants in an aqueous solution. If the solutes and surfactant are of the same charge type, the solutes will be expelled into the permeate by a process called ion-expulsion ultrafiltration (IEUF) (16). In terms of removal of metal ions, MEUF and IEUF have proven to be very effective. However, a shortcoming of these methods is the lack of selectivity in the removal of cations of the same charge (12,14, 15,17).

In ligand-modified micellar-enhanced ultrafiltration (LM-MEUF) an amphiphilic ligand is solubilized in the micelles (1). The chelating ligand is derivatized with a long

[4]Corresponding author

0097–6156/92/0509–0180$06.00/0

$(C_{12}\text{-}C_{16})$ hydrocarbon "tail" to promote solubilization into the micelle. The chelating moiety of the ligand is chosen to provide the desired complexation selectivity for the target cation(s). As shown schematically in Figure 1, the target ion, complexed by the solubilized ligand, should remain in the retentate during the filtration process. When used with a cationic surfactant, the overall separation will be improved due to the expulsion of uncomplexed metal ions into the permeate. Amphiphilic ligands have been used for separation of metal ions by extraction (*18*) and foam fractionation (*19-22*) techniques.

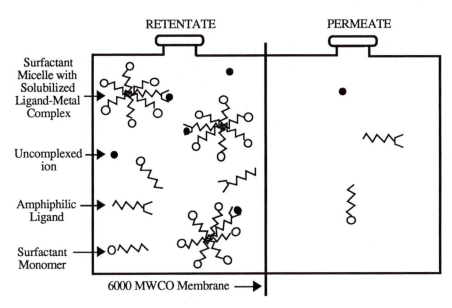

Figure 1. Schematic representation of a semi-equilibrium dialysis cell showing composition of retentate and permeate solutions after ~20 hrs. Metal ions complexed by the micelle-solubilized ligand are rejected by the membrane.

Stirred-cell LM-MEUF studies, using the ligand N-n-(dodecyl)-iminodiacetic acid (NIDA), showed rejection values as large as 99.2% for Cu^{2+}, with no rejection of Ca^{2+} (*1*). A model for the separation efficiency at equilibrium has been developed based on semi-equilibrium dialysis (SED) measurements (*7*).

Several criteria must be satisfied for a ligand to be useful in LM-MEUF. The ligand, and its metal-ion complex, should have large solubilization constants for the surfactant used. This condition must be satisfied in order to minimize loss of the amphiphilic ligand during the separation process. The solubility of the ligand, in absolute terms, in the micellar phase should be at least 1-2% of the total surfactant concentration in order to provide sufficient metal ion binding capacity. For typical surfactant concentrations used in this method (0.05-0.10 M), the total ligand solubility should be at least 0.5-1.0 mM. Moreover, the stoichiometry of the complex should be 1:1 (ligand:metal) to maximize the metal ion binding capacity per mole of solubilized ligand. The ligand should complex the target metal ion(s) quantitatively, i.e., $[ML] \geq 0.99[M]_{tot}$. This condition requires relatively high values of the effective complex formation constant, K_{ML}', for the conditions employed. For example, if the initial concentrations of the ligand and metal ion are both 1 mM, the value of K_{ML}' must be at least 1×10^7 M^{-1} to reduce the free metal ion concentration to one percent of its initial value. The ligand must have a high degree of complexation selectivity for the target cation relative to other ions in solution. For a system containing the ligand and a mixture of two cations, each initially at 1 mM, the complexation constant for the target cation (M) must be at least 10^6-times larger than that of the other cation (M^*) in order to satisfy the conditions: $[ML] \geq 0.99[M]_{tot}$ and $[M^*L] \leq 0.01[M^*]_{tot}$. The absolute (strength) and relative (selectivity) values of the complexation constants may be altered by appropriate ligand design. Structural factors that play a role in metal ion complexation include the number and type (O,N,S) of ligand donor atoms, chelate ring size (5- or 6-membered), and the geometry of the donor atom lattice (23).

The work reported here utilizes ligands containing only nitrogen (N) donor atoms. These compounds were synthesized by the condensation of a straight-chain alkyl bromide with either diethylenetriamine (DIEN) or di-2-picolylamine (DPA). The resulting ligands, N-(n-dodecyl)-diethylenetriamine (DDIEN), N-(n-dodecyl)-di-2-picolylamine (DDPA), and N-(n-hexadecyl)-di-2-picolylamine (HDPA), are:

DDIEN   DDPA   HDPA

The structural differences among these ligands allow evaluation of the effects of alkyl chain length and aliphatic vs. aromatic nitrogen donor atoms on the suitability of these compounds as ligands for metal ion separations using LM-MEUF. Comparison of results for DDPA (C_{12}) and HDPA (C_{16}) should reveal the effect of chain length on the solubilization of the ligand into the micelle. DIEN and DPA are both tridentate ligands that form linked, 5-membered chelate rings in the metal ion complex. However, the aliphatic nitrogen donor atoms in DIEN are more basic than the aromatic nitrogens of DPA (24). Therefore, the pH-dependent distribution of protonated and complexed species for the two ligands will be different. The protonation and complexation constants for DIEN and DPA in aqueous solution are listed in Table I. These values were used to calculate the pH-dependent distribution of the various ligand species in the absence and presence of Cu^{2+} (25).

This report gives results of equilibrium studies of the solubilization and metal-ion complexation behavior of three amphiphilic triamine ligands in micellar solutions. The separation selectivity has been evaluated using results of studies of solutions containing mixtures of Cu^{2+} and Ca^{2+}. The studies were carried out using the SED

Table I. Equilibrium Constants $(\log_{10} K_i)$ for DIEN and DPA[a]

Reaction	DIEN	DPA
$H^+ + L \rightleftharpoons HL^+$	9.84	7.29
$H^+ + HL^+ \rightleftharpoons H_2L^{2+}$	9.02	2.60
$H^+ + H_2L^2 \rightleftharpoons H_3L^{3+}$	4.23	1.12
$L + Cu^{2+} \rightleftharpoons CuL^{2+}$	15.9	14.4
$L + CuL^{2+} \rightleftharpoons CuL_2^{2+}$	5.0	4.6

[a] Equilibrium constants from reference 24.

technique, a method that has been shown to give results that predict the behavior of the retained species in an ultrafiltration system (3,4,7,26) and that is fast and economical. The studies employed surfactants with positive, neutral, and negative head groups in order to examine the role of charge on ligand solubilization and metal-ion complexation. These processes were studied as a function of $[Cu^{2+}]$, $[Ca^{2+}]$, pH, and in the presence and absence of added electrolyte (NaCl).

Experimental Section

Chemicals. N,N-dimethyldodecylamine-N-oxide (DDAO) and cetyl trimethyl-ammonium bromide (CTAB) were purchased from Fluka Chemical Corp. Cetyl pyridinium chloride (CPC) and polyoxyethylene nonyl phenyl ether, with an average degree of polymerization of 10 $(NP(EO)_{10})$, were obtained from Hexacel Corp. and GAF Corp., respectively. These surfactants were used without further purification. Sodium dodecylsulfate (SDS), obtained from Fisher Scientific Co., was purified by recrystallization from an ethanol-water mixture. 4-dodecyl-diethylenetriamine (DDIEN) was purchased from Eastman Kodak Co. Metal ion salts were all reagent grade. Standard pH buffers (Fisher Scientific Co.) and Hepes buffers (Sigma Chemical Co.) were used as provided.

N-(n-dodecyl)-di-2-picolylamine (DDPA) and N-(n-hexadecyl)-di-2-picolylamine (HDPA) were synthesized by condensation of the respective 1-bromoalkanes (Aldrich Chemical Co.) with di-2-picolylamine (Nepera Corp.) according to literature procedures (27). The crude products were purified using column chromatography over silica gel with chloroform containing increasing amounts of methanol (0-5%) as the eluent. The alkanes were used as obtained and the di-2-picolylamine was purified by vacuum distillation at 150°C (1 mm).

Procedures. Test solutions were prepared by combining known quantities of stock solutions of the desired components in a volumetric flask. Stock solutions of surfactants and NaCl were prepared by weight. Stock solutions of the copper and calcium salts were standardized by titration using primary standard EDTA (28). All solutions were prepared using doubly distilled water. The concentrations of the amphiphilic ligands were determined by spectrometric titration using a standardized copper solution as the titrant (29). Spectra were recorded using a Hitachi 100-80 double beam spectrophotometer. All test solutions were adjusted to the desired pH value using NaOH or HCl. pH measurements were made using a pH meter (Fisher-825MP or Corning-125) equipped with a combination electrode (Orion-8103).

The semi-equilibrium dialysis (SED) method has been used as described previously (5-8). The test solution is placed in the retentate compartment and pure water, or 0.10 M NaCl solution, is placed on the permeate side of the SED cell. The cells are equilibrated for 24 hours (30); then the final concentrations of Cu^{2+} and Ca^{2+} in the permeate and retentate compartments are analyzed. The metal ion concentrations are determined according to previously described procedures (1,7) using a Varian Spectr AA-20 atomic absorption spectrophotometer equipped with a graphite furnace.

Solubilization equilibrium constants for the ligands were determined at selected values of surfactant concentration, ligand concentration, and pH. These "conditional" equilibrium constants were determined from data obtained from SED experiments (31). The test solution initially placed in the retentate compartment contained surfactant and ligand at the desired concentrations and pH. After 24 hours, the concentration of the ligand in the permeate was measured by spectrometric titration with standard Cu^{2+}. Corrections were not made for the presence of surfactant micelles and solubilized ligand in the permeate compartment. The CMC of DDIEN was measured using a DuNouy ring tensiometer. Solutions of DDIEN were prepared at concentrations ranging from 5.0 x 10^{-5} M to 1.1 x 10^{-3} M at pH 7.0. The CMC (pH 7.0, 25°C) was determined from a plot of surface tension (10 replicates) versus DDIEN concentration.

Results and Discussion

Solubility of Ligands. The solubility of the ligands, in water and in surfactant solutions, is an important consideration in evaluating their usefulness in the LM-MEUF process. A reasonable level of solubility in the surfactant micelles is required to provide sufficient metal ion binding capacity to remove the target cations. Generally, the mole fraction of ligand in the micelle, X_L, should be about 0.01-0.02. Thus, ligand solubilities should be in the range 0.5-2.0 mM for the total surfactant concentrations employed in these studies (0.05-0.10 M). The amphiphilic ligands, DDIEN, DDPA, and HDPA, have solubilities ≥ 0.5 mM in the pH range 5.0-7.8 with the five surfactants tested in this study. More extensive studies with HDPA, in the pH range 1.9-9.2, indicate that its solubility exceeded 3.0 mM in 0.10 M SDS, 0.05 M CTAB, and 0.10 M DDAO. The solubilities in water of HDPA and DDPA are 0.7 mM (pH=6.0) and 290 mM (pH=5.2), respectively. The relatively high solubility of DDPA in water could lead to undesirable levels of ligand loss during the LM-MEUF process. The addition of four methylene groups to give HDPA, decreased the aqueous solubility of the ligand dramatically. Since both ligands would exist primarily in the monoprotonated form ($\geq 90\%$ HL$^+$, at pH ≤ 6.0) (24), the slightly higher pH value of the HDPA solution could not account for the large difference in solubility.

DDIEN shows a very high solubility in water at pH=7.0. At this pH value, where the ligand probably exists as an HL^+/H_2L^{2+} mixture (24), the solubility is at least 50 mM. In fact, surface tension studies show that DDIEN forms micelles with a CMC = 0.016 mM at 25°C (17). This value is consistent with surface tension-concentration data previously reported at unspecified values of temperature and pH (20). The ability of DDIEN to form micelles without additional surfactant suggests that this ligand could be used alone for MEUF separations, although the CMC of 0.16 mM would result in losses of ligand into the permeate stream.

Solubilization Constants of Ligands. In addition to the requirement that a reasonable amount of the ligand dissolves in the surfactant micelles ($X_L > 0.01$), the ligand should also partition strongly into the micellar phase. The partition, or solubilization constant, K_L, for the ligand is defined by the equation,

$$K_L = X_L/c_L = [L]_{mic}/\{c_L([L]_{mic} + [surf]_{mic})\} \qquad (1)$$

where X_L is the mole fraction of the ligand in the micelles, c_L is the molar concentration of free ligand in the bulk aqueous solution, and $[L]_{mic}$ and $[surf]_{mic}$ are the micellar concentrations of ligand and surfactant, respectively. Since the ligands may exist in a number of protonated forms (H_nL^{n+}, n=0-3), and the degree of solubilization may depend on the ratio, $[L]/[surf]$, the values of K_L reported are more accurately termed "conditional" solubilization constants, K_L', valid at the pH and $[L]/[surf]$ values specified. However, these conditional values, listed in Table II, should serve as a useful guide in the evaluation of ligands for LM-MEUF.

Table II.   Solubilization Constants for DDIEN, DDPA, and HDPA

Ligand[b]	pH[c]	K_L', M^{-1} [a]		
		CTAB(0.05 M)	DDAO(0.1 M)	SDS(0.1 M)
DDIEN(1 mM)	7.0	130	1150[d]	>2000
DDPA(0.5 mM)	7.0	90(20)	60(3)	40(7)
HDPA(0.5 mM)	5.0	660(90)[e]	690(30)	900(250)
	8.5	780(10)	270(50)	510(70)

[a] K_L' values at 25 °C for conditions listed; uncertainties, in parentheses, given as 1 standard deviation. [b] $[L]_{tot}$ given in parenthesis. [c] Average pH value ($\pm$ 0.2), measured in permeate. [d] $[DDAO]_{tot}$ = 0.050 M. [e] $[CTAB]_{tot}$ = 0.040 M.

Inspection of Table II reveals several trends for the values of the solubilization constants. For DDIEN at pH = 7.0, the values of K_L' increase dramatically as the surfactant is changed from CTAB to DDAO to SDS. At the pH specified, DDIEN most likely exists in the diprotonated form (H_2L^{2+}) in all three surfactants (*32-34*). Thus, electrostatic interactions appear to have a major effect on the observed value of K_L' for DDIEN. For DDPA at pH = 7.0, the ligand is most likely a mixture of the unprotonated and monoprotonated forms, with the percentage of the neutral form decreasing as the charge of the surfactant decreases from CTAB (+1) to SDS (-1) (*32-34*). The values of K_L' vary by less than a factor of two among the three surfactants tested, decreasing slightly going from CTAB to SDS. At pH = 7.0, DDAO exists predominantly (~90%) in the neutral, zwitterionic form (*35*). Comparison of the K_L' values of DDIEN and DDPA, particularly in DDAO where electrostatic interactions should be minimal, indicates that DDIEN is much more soluble in these surfactants than DDPA. The trend in K_L' values for DDPA indicates that electrostatic contributions are minimal at the pH specified. Comparison of DDPA and HDPA reveals that the K_L' values for the latter are 5-20 times larger than those for DDPA.

Since the chelating moieties are identical this change must be a result of the addition of four methylene units to the hydrocarbon tail. As previously found for DDPA, the values of $K_L{}'$ vary by less than a factor of three among the surfactants tested in the pH range 5.0-8.5. At the lower pH value, where HDPA is more extensively protonated, $K_L{}'$ varies, albeit slightly, as expected. At pH = 8.5, however, no apparent trend is discernable. Under the conditions employed, the values of $K_L{}'$ for DDIEN in DDAO and SDS and for HDPA in each of the surfactants are sufficiently large so that at least 97% of the total ligand concentration is solubilized in the micelles.

SED Complexation Studies. The complexation behavior of DDIEN and HDPA was evaluated using the SED method. DDPA was not evaluated further because the low values of $K_L{}'$ for the compound indicate that there would be excessive loss of the amphiphilic ligand, particularly with positively charged surfactants like CTAB and CPC. For the ligands tested, the initial concentrations of Cu^{2+} and, in some cases, Ca^{2+}, were varied at a fixed concentrations of surfactant, ligand, and pH. For HDPA in CPC, the effect of added electrolyte (0.1 M NaCl) was evaluated. The rejection coefficient for Cu^{2+}, R_{Cu}, was calculated using the following equation,

$$R_{Cu} = \{\ 1 - [Cu^{2+}]_p/[Cu^{2+}]_r\} \times 100\% \tag{2}$$

where $[Cu^{2+}]_p$ and $[Cu^{2+}]_r$ are the measured concentrations of copper ion in the permeate and retentate solutions, respectively.

No results are listed for studies using 0.05 M DDIEN at pH = 7.0 without additional surfactant. For solutions containing 1.1 ± 0.1 mM Ca^{2+} and with $[Cu^{2+}]_0$ ranging from 0.001 to 0.64 mM, no rejection of Cu^{2+} was observed. Likewise, no rejection of copper was found for 0.50 mM DDIEN in 0.05 M CTAB under similar conditions. Separate SED studies were carried out using 0.5 mM DDIEN and 0.5 mM HDPA in 0.10 M SDS solutions containing 1.1-1.2 mM Ca^{2+} and various concentrations of Cu^{2+}. In both cases the values of R_{Cu} were greater than 98.4%, even for solutions whose initial copper:ligand ratios, $[Cu^{2+}]_0$:$[ligand]_0$, were greater than unity, the point of ligand saturation. The latter results were not surprising in view of the the non-selective rejection of metal ions by solutions containing SDS micelles (*12,14,15,17*). Results of the SED studies of DDIEN with CPC and NP(EO)$_{10}$ and of HDPA with CPC, CTAB, DDAO and NP(EO)$_{10}$ are listed in Tables III-V.

SED Studies with DDIEN. Table III lists the results for SED studies of DDIEN with CPC, a positively charged surfactant. At pH = 7.0 and in the absence of added electrolyte, DDIEN performs poorly, with R_{Cu} values no greater than 67% for $[Cu^{2+}]_0/[DDIEN]_0$ ratios less than unity. The presence of 0.10 M NaCl in the system results in an increase of R_{Cu} to ~96% for the same $[Cu^{2+}]_0/[DDIEN]_0$ range, with no rejection for Ca^{2+}. In the former case, the affinity of the ligand for Cu^{2+} is not sufficient to overcome the electrostatic interaction between the divalent copper ion and the net positive charge of the micelle surface. In fact, an ion expulsion effect is observed, wherein the final concentrations of Cu^{2+} and Ca^{2+} are greater in the permeate than in the retentate solutions (*16*). The addition of 0.10 M NaCl raises the

Table III. Semi-Equilibrium Dialysis Results for DDIEN in CPC and NP(EO)$_{10}$

CPC: [CPC]$_0$ = 0.10 M, [NaCl]$_0$ = 0.0 M

Initial Ratio [a]		Retentate(mM)[b]		Permeate(mM)[b]		Cu Rejection(%)
Cu:L	Ca:L	[Cu]	[Ca]	[Cu]	[Ca]	
0.390	2.66	0.106	0.200	0.035	0.952	66.7
0.460	2.52	0.092	0.136	0.042	0.716	54.3
0.960	2.50	0.144	0.160	0.234	0.700	----
1.409	2.52	0.212	0.092	0.328	0.901	----
1.636	2.08	0.216	0.064	0.565	0.688	----
0.894	2.94	0.129	0.088	0.222	0.265	----
0.960	0.94	0.144	0.030	0.258	0.270	----
0.870	0.03	0.108	0.010	0.225	0.046	----
0.002	2.28	0.001	0.032	0.001	0.888	----

[CPC]$_0$ = 0.10 M, [NaCl]$_0$ = 0.10 M

Cu:L	Ca:L	[Cu]	[Ca]	[Cu]	[Ca]	
0.472	2.20	0.221	0.746	0.008	0.970	96.4
0.818	2.10	0.367	0.532	0.015	0.761	95.9
1.428	2.22	0.576	0.652	0.104	0.965	81.9
1.828	2.22	0.615	0.529	0.203	0.816	67.5

NP(EO)$_{10}$: [NP(EO)$_{10}$]$_0$ = 0.10 M, [NaCl]$_0$ = 0.0 M

Initial Ratio[a]		Retentate(mM)[b]		Permeate(mM)[b]		Cu Rejection(%)
Cu:L	Ca:L	[Cu]	[Ca]	[Cu]	[Ca]	
0.392	2.44	0.152	0.468	0.011	0.591	92.8
0.728	2.46	0.264	0.498	0.045	0.594	83.0
1.056	2.36	0.312	0.456	0.054	0.525	82.7
1.280	2.60	0.378	0.537	0.063	0.660	83.3
1.500	2.60	0.474	0.558	0.072	0.672	84.8
0.984	5.12	0.321	1.03	0.064	1.12	80.1
0.888	1.42	0.246	0.164	0.036	0.266	85.4
0.792	0.03	0.234	0.019	0.034	0.029	85.5
0.006	2.44	0.003	0.498	0.000	0.558	----

[NP(EO)$_{10}$]$_0$ = 0.10 M, [NaCl]$_0$ = 0.10 M

Cu:L	Ca:L	[Cu]	[Ca]	[Cu]	[Ca]	
0.648	2.22	0.316	0.679	0.010	0.760	96.8
1.092	2.10	0.493	0.654	0.038	0.745	92.3
1.304	2.04	0.531	0.635	0.075	0.730	82.1
1.418	2.20	0.577	0.576	0.095	0.737	83.5

[a] [DDIEN]$_0$ = 0.50 mM, pH = 7.0

[b] Initial ratio of the metal-ligand concentrations, [M]$_0$:[DDIEN]$_0$.

NOTE: Each value is average of two replicates, average deviation < 5%.

ionic strength of the system and diminishes the CPC-metal ion repulsion (7) The decrease in R_{Cu} for $[Cu^{2+}]_0/[DDIEN]_0$ ratios greater than one indicates a 1:1 metal-ligand stoichiometry for the micellar Cu^{2+}-DDIEN complex under these conditions. SED runs using the neutral surfactant, $NP(EO)_{10}$, show no rejection of calcium ions and give higher rejection values for Cu^{2+} than those obtained using CPC. However, the ion expulsion effect is lost and the copper rejection values fall to ~83% at $[Cu^{2+}]_0/[DDIEN]_0$ ratios less than unity. Addition of 0.10 M NaCl to the system increases R_{Cu} when $[Cu^{2+}]_0 \leq [DDIEN]_0$. In solutions containing $NP(EO)_{10}$, the increased ionic strength screens the electrostatic repulsion between Cu^{2+} and the mono- and diprotonated forms of micellar DDIEN which most likely predominate at pH 7.0 (24,33-35). The relationship between R_{Cu} and $[Cu^{2+}]_0/[DDIEN]_0$ suggests 1:1 stoichiometry for the micellar copper-DDIEN complex in $NP(EO)_{10}$.

SED Studies with HDPA. Table IV lists the results of SED runs with CPC in the absence and presence of 0.10 M NaCl. The results of additional studies without 0.10 M NaCl, using the surfactants CTAB, DDAO, and $NP(EO)_{10}$, are listed in Table V.

SED Studies with HDPA in CPC. Trial runs were carried out at pH = 6.6($\pm$0.3) in the absence and presence of 0.10 M NaCl. For the experiments without added NaCl, the rejection of copper at constant calcium ($[Ca^{2+}]_0 = 1.2\pm0.1$ mM) varied from 85.2-99.6% for solutions where $[Cu^{2+}]_0 \leq [HDPA]_0$. At a fixed value of $[Cu^{2+}]_0$, (0.57$\pm$0.09 mM), variation of $[Ca^{2+}]_0$ from 0.047 to 4.66 mM had little effect, with R_{Cu} values in the range 99.0-99.6%. No rejection of Ca^{2+} was observed. In fact, the ion expulsion effect resulted in $[Ca^{2+}]_p/[Ca^{2+}]_r$ (permeate/retentate) ratios of ~10-20. In the presence of 0.10 M NaCl the ion expulsion effect disappears, with $[Ca^{2+}]_p/[Ca^{2+}]_r$ ratios falling to ~1.0. The rejection of copper improves at low concentrations, with R_{Cu} values from 98.3-99.8% for $[Cu^{2+}]_0/[HDPA]_0$ ratios less than one. These effects are clearly illustrated by the plots shown in Figure 2. The increased rejection values at low concentrations of copper ion are a result of the diminished ion expulsion effect on Cu^{2+} in the presence of excess NaCl (7). As shown in Table IV and Figure 2, the copper rejection decreases significantly when $[Cu^{2+}]_0 > [HDPA]_0$, indicating that the copper-HDPA complexes in CPC have a 1:1 metal ion-ligand stoichiometry.

HDPA with CTAB. Results from SED studies of 0.62 mM HDPA in 0.04 M CTAB at pH = 7.5$\pm$0.5 are listed in Table V. For runs where $[Cu^{2+}]_0/[HDPA]_0$ <0.8 and where $[Ca^{2+}]_0 = 0.64\pm0.02$ mM, the rejection of copper is in the range 98.6-99.8%, with no rejection of Ca^{2+}. Since CTAB is a positively charged surfactant, Ca^{2+} ion is concentrated in the permeate, with $[Ca^{2+}]_p/[Ca^{2+}]_r$ ratios in the range 5-17. The dependence of R_{Cu} on the copper-ligand ratio suggests that both 1:1 and 1:2 metal-ligand complexes are formed in CTAB micelles. The higher rejection values at low copper concentrations, compared with results using CPC, may be due more to favorable complexation at the higher pH employed in the CTAB study or to better solubilization of the ligand and its copper complex in CTAB relative to CPC.

HDPA with DDAO. DDAO has an ionizable head group with pKa ~5 (*35*), and exists as a positively charged surfactant at low pH and as a neutral, zwitterionic surfactant for pH values above eight. At the pH employed (5.5) for the study reported, about 35% of the surfactant is in the protonated form, and DDAO does not bind Cu^{2+} ions to any measurable extent (*17*). As the $[Cu^{2+}]_0/[HDPA]_0$ ratios increase from 0.1 to 1.0, the copper rejection increases from 88.6-96.7% with no rejection of Ca^{2+}. For [Cu]/[L] ratios greater than one, R_{Cu} decreases abruptly, indicating that complexes with 1:1 stoichiometry predominate. The partial positive charge on the DDAO micelles gives rise to a mild ion expulsion effect, with $[Ca^{2+}]_p/[Ca^{2+}]_r = 4.5 \pm 0.5$.

Table IV.   Semi-Equilibrium Dialysis Results for HDPA in CPC[a]

[NaCl] = 0.0 M.

Initial Ratio[b]		Retentate(mM)[c]		Permeate(mM)[c]		Cu Rejection(%)
Cu:L	Ca:L	[Cu]	[Ca]	[Cu]	[Ca]	
0.390	2.23	0.132	0.064	0.0196	1.070	85.2
0.700	1.94	0.278	0.074	0.0021	1.224	99.2
0.861	7.52	0.382	1.725	0.0016	4.700	99.6
0.960	2.24	0.373	0.104	0.0037	1.225	99.0
1.087	0.84	0.284	0.052	0.1690	0.491	40.5
1.513	1.87	0.406	0.126	0.3140	1.240	22.7
0.763	0.08	0.298	0.020	0.0021	0.077	99.3
0.015	1.34	<.005	0.044	0.0015	0.822	----
0.--[d]	1.60	0.016	0.074	0.3810	1.010	----

[NaCl] = 0.10 M.

Initial Ratio[b]		Retentate(mM)[c]		Permeate(mM)[c]		Cu Rejection(%)
Cu:L	Ca:L	[Cu]	[Ca]	[Cu]	[Ca]	
0.190	1.92	0.100	0.464	0.0004	0.445	99.6
0.295	1.53	0.156	0.383	0.0026	0.435	98.3
0.519	1.41	0.289	0.352	0.0006	0.359	99.8
0.656	1.18	0.378	0.348	0.0008	0.343	99.8
1.269	1.22	0.519	0.229	0.0421	0.242	91.9
0.750	0.15	0.436	0.086	0.0038	0.074	99.1
0.--[d]	1.20	0.046	0.312	0.1075	0.369	----

[a] $[CPC]_0 = 0.10$ M, $[HDPA]_0 = 0.62$ mM, pH = 6.6 ($\pm$ 0.1).

[b] Initial ratio of the metal-ligand concentrations, $[M]_0:[DDIEN]_0$.

[c] Each value is average of two replicates. [d] No HDPA present.

Table V.   Semi-Equilibrium Dialysis Results for HDPA in CTAB, DDAO, and $NP(EO)_{10}$

CTAB: $[CTAB]_0 = 0.04$ M, $[HDPA]_0 = 0.62$ mM

Initial Ratio[a]		Retentate(mM)[b]		Permeate(mM)[b]		Cu Rejection(%)
Cu:L	Ca:L	[Cu]	[Ca]	[Cu]	[Ca]	
0.072	1.04	0.032	0.059	0.0002	0.574	99.4
0.145	1.10	0.088	0.034^d	0.0004	0.568	99.5
0.295	1.03	0.155	0.060^d	0.0006	0.532	99.6
0.402	0.99	0.203	0.060	0.0009	0.568	99.6
0.502	1.00	0.238	0.070	0.0006	0.588	99.7
0.629	1.00	0.274	0.071	0.0039	0.585	98.6
0.687	1.00	0.269	0.058	0.0028	0.497	99.0
0.866	1.06	0.290	0.096	0.0760	0.495^d	73.8
1.379	1.05	0.342	0.159	0.1695	0.544	50.4
$0.--^c$	0.30	<.005	0.060	0.134^c	0.174^d	----

DDAO: $[DDAO]_0 = 0.10$ M, $[HDPA]_0 = 0.50$ mM

Initial Ratio[a]		Retentate(mM)[b]		Permeate(mM)[b]		Cu Rejection(%)
Cu:L	Ca:L	[Cu]	[Ca]	[Cu]	[Ca]	
0.092	2.78	0.068	0.236	0.008	0.922	88.2
0.152	2.37	0.090	0.170	0.008	0.852	91.1
0.330	2.37	0.122	0.151	0.011	0.840	90.9
0.638	2.41	0.274	0.178	0.015	0.812	94.5
1.108	2.50	0.482	0.180	0.016	0.847	96.7
1.560	2.79	0.506	0.186	0.058	0.843	88.5
2.480	2.52	0.709	0.224	0.338	0.868	52.3
0.636	0.02	0.262	0.076	0.008	0.123	96.9
0.032	2.61	0.004	0.196	0.002	0.909	----

$NP(EO)_{10}$: $[NP(EO)_{10}]_0 = 0.050$ M, $[HDPA]_0 = 0.50$ mM

Initial Ratio[a]		Retentate(mM)[b]		Permeate(mM)[b]		Cu Rejection(%)
Cu:L	Ca:L	[Cu]	[Ca]	[Cu]	[Ca]	
0.246	2.03	0.076	0.446	0.0108	0.396	85.8
0.408	2.03	0.116	0.512	0.0036	0.466	96.9
0.562	1.98	0.192	0.460	0.0032	0.480	98.3
0.718	1.80	0.299	0.427	0.0034	0.521	98.9
0.788	2.00	0.320	0.415	0.0230	0.448	92.8
1.176	1.94	0.396	0.444	0.1135	0.476	71.3
1.300	1.92	0.439	0.462	0.2058	0.502	53.1

[a] Initial ratio of the metal-ligand concentrations, $[M]_0:[DDIEN]_0$.

[b] Each value is average of two replicates. [c] No HDPA present.

[d] Only one value obtained.

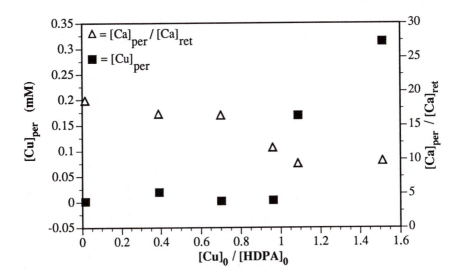

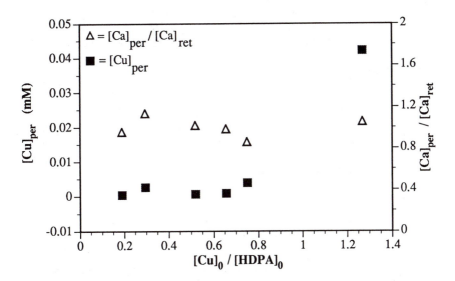

Figure 2. Plots of the ratio, $[Ca^{2+}]$(permeate):$[Ca^{2+}]$(retentate), and $[Cu^{2+}]$ (permeate) vs. $[Cu^{2+}]$(initial)/[HDPA](initial) for SED runs of HDPA in CPC; [CPC] = 0.1 M, [HDPA] = 0.62 mM, $[Ca^{2+}]$ = 1.0 mM Top; values for system with no added NaCl. Bottom; values for system containing 0.10 M NaCl.

HDPA with NP(EO)$_{10}$. The final entries in Table V list the results for SED studies with HDPA at pH = 4.5 in solutions containing 0.98 mM Ca^{2+} and 0.050 M NP(EO)$_{10}$. For solutions with $[Cu^{2+}]_0/[HDPA]_0 < 1$, the rejection values for copper vary from 85.8 to 98.9%. Since NP(EO)$_{10}$ is neutral, any ion expulsion effects would arise from the interaction between solubilized $H_n HDPA^{n+}$ and Cu^{2+}. The finding that $[Ca^{2+}]_p/[Ca^{2+}]_r = 1.0\pm0.1$ indicates that such interactions are minimal. As found for other surfactants, Cu^{2+} forms predominantly 1:1 complexes with solubilized HDPA.

Conclusions

Each of the alkyl triamine ligands has sufficient solubility ($X_L \geq 0.01$) in the surfactants tested to provide adequate metal chelating capacity. However, the relatively small values of K_L' found for DDPA, limit its effectiveness. With DDIEN and HDPA, at least 97% of the ligand is partitioned into the micelles. Both DDIEN and HDPA form metal-ligand complexes with predominantly 1:1 stoichiometry in the surfactants tested. In mixtures containing Cu^{2+} and Ca^{2+}, HDPA provides excellent rejection of Cu^{2+} with no rejection of Ca^{2+}. With the cationic surfactants CPC and CTAB, the uncomplexed Ca^{2+} ions are concentrated in the permeate solution, providing a more effective <u>overall</u> separation than found with neutral surfactants. DDIEN exhibits marginal Cu^{2+} rejection at neutral pH in the absence of added NaCl. In this pH range the higher degree of protonation of DDIEN reduces its effective complexation strength relative to HDPA. The addition of excess NaCl to the systems containing CPC screens the electrostatic repulsion between the surfactant and the metal ions, resulting in higher rejection values for Cu^{2+}, but decreased expulsion of unbound Ca^{2+} to the permeate.

Acknowledgments

Financial support for this research was provided by the National Science Foundation (Grant CBT-8814147), the Department of Energy Office of Basic Energy Sciences (Grant DE-FG05-84ER13678), the Bureau of Mines, the Department of Energy, the Environmental Protection Agency, the Oklahoma Mining and Minerals Resources Research Institute, Aqualon Corp., Sandoz Corp., E. I. du Pont de Nemours & Co., Kerr-McGee Corp., and Union Carbide Corp. The gift of di-2-picolylamine from Nepera Corp. is gratefully acknowledged.

Literature Cited

1. Klepec, J.; Simmons, D. L.; Taylor, R. W.; Scamehorn, J. F.; Christian, S. D. *Sep. Sci.. Technol.* **1991**, *26*, 165-173.
2. Leung, P. S. in *Ultrafiltration Membranes and Applications*; Cooper, A. R., Ed.; Plenum Press: New York, NY, 1979, p. 415.
3. Dunn, R. O.; Scamehorn, J. F.; Christian, S. D. *Sep. Sci. Technol.* **1985**, *20*, 257-284.
4. Dunn, R. O.; Scamehorn, J. F.; Christian, S. D. *Sep. Sci. Technol.*, **1987**, *22*, 763-789.
5. Gibbs, L. L.; Scamehorn, J. F.; Christian, S. D. *J. Membr. Sci.* **1987**, *30*, 67-74.
6. Bhat, S. N.; Smith, G. A.; Tucker, E. E.; Christian, S. D.; Smith, W.; Scamehorn, J. F. *Ind. Eng. Chem. Res.* **1987**, *26*, 1217-1222.

7. Dharmawardana, U.; Christian, S. D.; Taylor, R. W.; Scamehorn. J. F. *Langmuir*, in press.
8. Smith, G. A.; Christian, S. D.; Tucker, E.E.; Scamehorn, J. F. in *Use of Ordered Media in Chemical Separations*; Hinze, W. L.; Armstrong, D. W., Eds.; ACS Symposium Series 342; American Chemical Society: Washington DC, 1987; pp. 184-198.
9. Christian, S. D.; Scamehorn, J. F. in *Surfactant-Based Separation Processes;* Scamehorn, J. F.; Harwell, J. H., Eds.; Dekker: New York, NY, 1989, Chap. 1.
10. Scamehorn, J. F.; Harwell, J. H. in *Surfactants in Chemical Process Engineering;* Scamehorn, J. F.; Harwell, J. H., Eds.; Dekker, New York, NY, 1988, p. 77-125.
11. Christian, S. D.; Tucker, E. E.; Scamehorn, J. F. *Am. Environ. Lab.* **1990**, (February), 13-20.
12. Scamehorn, J. F.; Christian, S. D.; Ellington, R. T. in *Surfactant-Based Separation Processes;* Scamehorn, J. F.; Harwell, J. H., Eds.; Dekker: New York, NY, 1989, Chap. 2.
13. Christian, S. D.; Bhat, S. N.; Tucker, E. E.; Scamehorn, J. F.; El-Sayed, D. A. *AIChE J.* **1988**, *34*, 189-194.
14. Scamehorn, J. F.; Ellington, R. T.; Christian, S. D.; Penney, B. W.; Dunn, R. O.; Bhat, S. N. in *Recent Advances in Separation Techniques - III;* Li, N. N. Ed.; AIChE Symposium Series; **1986**, *250*, 48-58.
15. Dunn, R. O.; Scamehorn, J. F.; Christian, S. D. *Colloids Surf.* **1989**, *35*, 49-56.
16. Christian, S. D.; Tucker, E. E.; Scamehorn, J. F.; Lee, B. H.; Sasaki, K. J. *Langmuir* **1989**, *5*, 876-879.
17. Schovanec, A. L. Ph.D. Thesis, University of Oklahoma, Norman, OK, 1991.
18. Hinze, W. L. in *Use of Ordered Media in Chemical Separations*; Hinze, W.L.; Armstrong, D.W., Eds.; ACS Symposium Series 342; American Chemical Society: Washington DC, 1987; p. 1-82.
19. Okamoto, Y.; Chou, E. J. *Separ. Sci. Technol.* **1975**, *10*(6), 741-753.
20. Okamoto, Y.; Chou, E. J. *Sep. Sci. Technol.* **1978**, *13*(5), 439-448.
21. Chou, E. J.; Okamoto, Y. *J. Water Pollut. Control Fed.* **1976**, *48*, 2747-2753.
22. Schnepf, R. W.; Gaden, E. L, Jr.; Mirocznok, E. Y.; Schonfeld, E. *Chem. Eng. Progr.* **1959**, *55*(5), 42-46.
23. Lehn, J. M. *Struct. Bonding(Berlin)* **1973**, *16*, 1-69.
24. Smith, R. M.; Martell, A. E. in *Critical Stability Constants*; Smith, R. M.; Martell, A. E., Eds.; Volume 2; Plenum Press: New York, NY, 1975.
25. Perrin, D. D.; Sayce, I. G.; *Talanta*; **1967**, *14*, 833-842.
26. Christian, S. D.; Tucker, E. E. *Am. Lab.*(Fairfield, Conn.) **1982**, *14*(8), 36.
27. Stein, A. S.; Gregor, A. P.; Spoerri, P. E. *J. Am. Chem. Soc.* **1955**, *77*, 191-192.
28. Vogel, A. I. *Quantitative Inorganic Analysis*, 3rd Ed.; Wiley: New York, NY, 1961; p. 441.
29. Polster, J.; Lachmann, H. *Spectrometric Titrations*; VCH Publishers: New York, NY, 1989, Chap. 13.
30. Christian, S. D.; Smith, G. A.; Tucker, E. E.; Scamehorn, J. F. *Langmuir* **1985**, *1*, 564-567.
31. Higazy, W. S.; Mahmoud, F. Z.; Taha, A. A.; Christian, S. D. *J. Soln. Chem.* **1988**, *17*, 191-202.
32. Shukla, S. S.; Meites, L. *Anal. Chim. Acta* **1985**, *174*, 225-235.
33. Underwood, A. L. *Anal. Chim. Acta* **1982**, *140*, 89-97.
34. Fernandez, M. S.; Fromherz, P. *J. Phys. Chem.* **1977**, *81*, 1755-1761.
35. Rathman, J. F.; Christian, S. D. *Langmuir* **1990**, *6*, 391-395.

RECEIVED January 24, 1992

Chapter 14

Polymeric Ligands for Ionic and Molecular Separations

S. D. Alexandratos, D. W. Crick, D. R. Quillen, and C. E. Grady

Department of Chemistry, University of Tennessee, Knoxville, TN 37996

An understanding of ligand-substrate interactions is important to the design of polymer
-supported reagents for the selective complexation of a single species in a multi-component solution. Such polymers may be applied
to the environmental separation of ions and
molecules as well as in chemical sensors and
as novel catalysts. A series of polymers has
been synthesized where substrate selectivity
arises through the polymer's bifunctionality.
The **dual mechanism bifunctional polymers** are
described and applied as both polymer beads
and membranes for the selective complexation
of metal ions and molecules.

The development of methodologies for the covalent
bonding of ligands to polymeric supports is important
for the preparation of novel reagents to be used in
ionic and molecular separations which are important to
the environment and for the preparation of chemical
sensors which can be used in the monitoring of environmentally sensitive areas. Our research focuses on
identifying the mechanisms by which immobilized ligands

0097–6156/92/0509–0194$06.00/0
© 1992 American Chemical Society

can interact selectively with targeted substrates and then synthesizing the corresponding polymer-supported reagents.

The selective complexation of ions and molecules by different reagents is best understood within the context of host-guest chemistry (1). Crown ethers discriminate among metal ions based on their ionic diameter (2), cryptands are still more selective (3) due to a higher degree of pre-organization (4), and cyclodextrins are shape selective for molecular substrates (5). Template polymerization is one method of pre-organizing the ligands on a polymer for greater substrate selectivity (6). Polymer-supported crown ethers (7) and cryptands (8) attempt to combine the selectivities inherent to the ligand with the advantages of using polymeric reagents, including ease of recovery and adaptability to continuous processes (9). Reactive polymers (10,11) form a unique category of reagents which influence the selectivities of ion exchange resins by superimposing an additional reaction (such as chelation) on the exchange mechanism. The dominant free energy of the superimposed reaction then directs the selectivity towards a targeted ion.

Metal Ion Complexation Reactions by Polymer-Supported Reagents

Our approach towards the preparation of highly selective reagents is to synthesize bifunctional cross-linked polymers and study cooperating ligand pairs with greater selectivities than the corresponding monofunctional reagents. These **dual mechanism bifunctional polymers** (DMBPs) have been divided into three classes, each of which displays a different recognition mechanism for metal ions (12).

The Class I DMBPs superimpose a reduction reaction on the ion exchange mechanism (13). The polymer with primary and secondary phosphinic acid ligands (Figure 1) has been studied extensively. The primary phosphinic acid groups will reduce metal ions, with highest selectivity for the mercuric ion (equation 1). The reduction potential of Cu(II) is at the lower limit of the polymer's ability to reduce ions to the free metal. The Cu(II) reduction approaches completion over a period of 240 hours while the Hg(II) reduction occurs over a few hours at room temperature. Increasing the temperature leads to a large increase in the rate of reduction. Optimization of the kinetics will also require a study of the effect of the polymer's macroporosity.

The Class II DMBPs are able to ion exchange with one ligand and coordinate with a different ligand. The phosphonate monoester/diester polymer (Figure 1) is a representative example and will be discussed in more detail below (14). The phosphonic acid/dimethylamine polymer is another example of this class.

The Class III DMBPs (Figure 1) combine ion exchange with a precipitation reaction, the latter arising from a quaternary amine ligand whose associated anion can be varied according to which cation is to be precipitated (15). Proper choice of the anion and alkyl groups on the amine leads exclusively to intra - particle precipitation (equation 2). Thus, for example, a bifunctional polymer with tributylammonium ion ligands coupled with the thiocyanate anion, when contacted with a 0.01N AgNO$_3$ solution, leads to precipitation of the AgSCN within the polymer matrix (70% sorption after a 17 h stir). Exchanging the polymer with KIO$_3$ yields a reagent specific for Pb(II) as iodate becomes the associated anion and Pb(IO$_3$)$_2$ precipitates.

Supported Ligand Synergistic Interaction. The complexation of metal ions by a pair of soluble complexing agents with one operating by ion exchange and the other

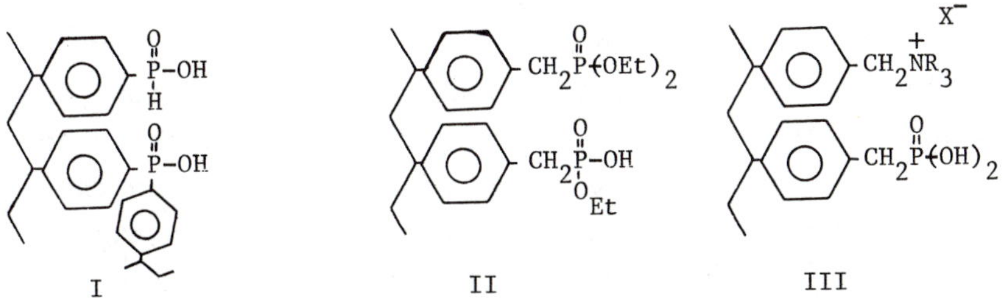

Figure 1. Chemical structures of the Class I, II, and III dual mechanism bifunctional polymers.

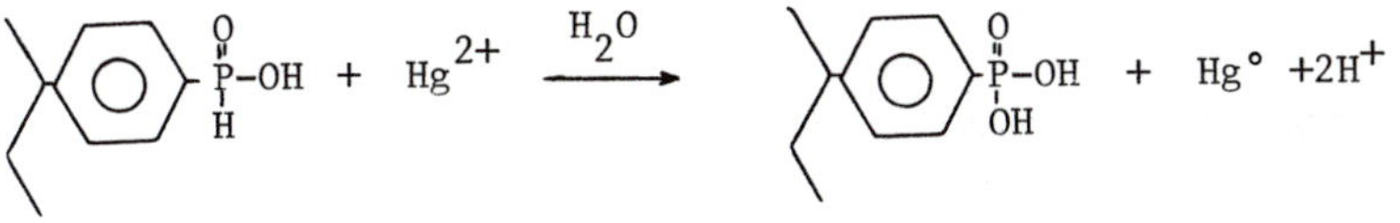

$$\text{(polymer)}-\overset{O}{\underset{H}{\overset{\|}{P}}}-OH + Hg^{2+} \xrightarrow{H_2O} \text{(polymer)}-\overset{O}{\underset{OH}{\overset{\|}{P}}}-OH + Hg^{\circ} + 2H^{+} \quad (1)$$

$$\text{(polymer)}\begin{matrix} CH_2NR_3^{+} X^{-} \\ CH_2\overset{O}{\underset{OH}{\overset{\|}{P}}}-OH \end{matrix} + M^{+} \longrightarrow \text{(polymer)}\begin{matrix} CH_2NR_3^{+} \\ CH_2\overset{O}{\underset{OH}{\overset{\|}{P}}}-O^{-} \end{matrix} + MX \downarrow \quad (2)$$

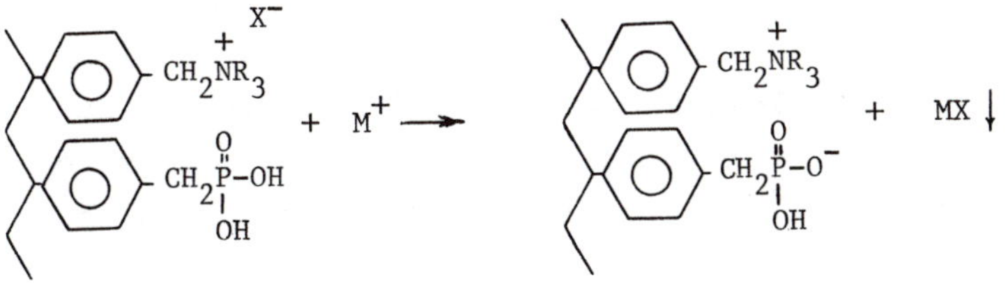

by coordination and in such a way that both groups complex more metal ion than either could alone, is well-known in solvent extraction chemistry (16). We first synthesized the phosphonate monoester/ diester polymer as well as the corresponding monofunctional polymers in order to define a polymeric analogue to this synergistic enhancement. In studying the complexation of Fe(III), Hg(II), Ag(I), Mn(II), and Zn(II) from acidic solutions (4.0 to 0.2N HNO_3) at a constant 4N nitrate background by the addition of varying levels of sodium nitrate, it is found that the behavior of the bifunctional polymer is similar to that found for the monofunctional resins for all ions except silver. Thus, from a background solution of 4N HNO_3, the distribution coefficients for Ag(I) are 439, 463, and 2924 for the monoester, diester, and bifunctional resins, respectively. On the other hand, the values for Hg(II) are 23, 3, and 15, in the order given above. The reason for a **supported** **ligand** **synergistic** **interaction** with Ag(I) has yet to be determined but is probably rationalized by a similar polarizability between the metal ion and the phosphoryl oxygen ligands.

A strong coordinative interaction is one mechanism by which reactivity controls recognition. Additional strong ion/ligand interaction mechanisms which can be used to control recognition are reduction and precipitation. Steric hindrance can also be used as another method of controlling recognition by limiting access to the coordinating sites. This seems to be true with the purely coordinating diester resins which display a selectivity depending upon the bulkiness of the ester moieties.

Polyfunctional Membranes via Interpenetrating Polymer Networks: Application to Metal Ion Separations

Polypropylene membranes which have been modified by sorbing organic solutions of complexing agents into the porous structure, have been utilized for highly selective metal ion transport from aqueous solutions (17). These supported liquid membranes (SLMs) are important to the separation of a wide variety of metal ions due to their versatility: the selectivity of the membrane is dependent on the complexing agent sorbed within the polypropylene and this can obviously be changed depending upon the targeted ion. A critical problem with the SLMs is that of stability: metal ion transport will eventually cease as the complexing agent and/or its attendant solvent dissolve out into the aqueous phase.

Our research with the SLMs has centered on the synthesis of highly stable membranes for long term metal ion transport. Since it is advantageous to maintain polypropylene as the basic support structure due to its commercial availability, a versatile membrane modification technique has been found to be that involving interpenetrating polymer networks [IPNs] (18). In this technique, a polymerizable metal ion complexing agent in a hydrophobic solvent is sorbed by the polypropylene and polymerized with a free radical initiator in the membrane either with or without additional comonomers (19). Stability studies are carried out by determining the membrane's ability to transport Cu(II) from pH 4.2 sulfate/bisulfate buffer solutions over a number of weeks (20). The concept is illustrated in Figure 2.

Di(undecenyl)phosphoric acid, $[CH_2=CH(CH_2)_9O]_2P(O)OH$, (DUP) was found to have the necessary properties needed for a complexing agent to be polymerized within a membrane: the ion exchange group is separated by a spacer

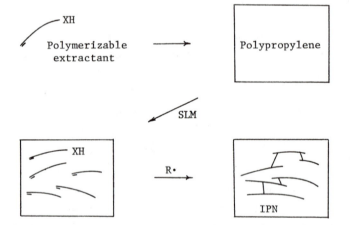

Figure 2. Highly stable polypropylene membranes through interpenetrating polymer network synthesis.

group from the vinyl moiety allowing it to remain mobile after polymerization; and the $-CH_2-$ adjacent to the vinyl group leads to chain transfer during the free radical polymerization (in the manner expected of allyl groups) which limits the degree of polymerization thus maintaining ligand mobility.

One-Component IPNs. A SLM of 0.4M di(2-ethylhexyl)-phosphoric acid (DEHP) in dodecane was studied as the control against which the performance of DUP membranes could be measured. It was found to have a permeability coefficient of $4x10^{-4}$ cm/sec and a lifetime of 45 days (extrapolated from its performance over 14 days). Since earlier studies had established that a minimum concentration of 1M DUP is required in order to prevent disentanglement of the polymerized product from the initial polymer network (21), SLMs at that DUP concentration were prepared. Interestingly, the 1M DUP IPN membrane had a shorter lifetime than the DEHP membrane, though its initial permeability, $2.5x10^{-4}$ cm/sec, indicated a transport rate comparable to the more mobile DEHP. Either IPN formation with 1M DUP did not produce sufficient entanglement with the polypropylene network, or the concentration of phosphorus acid moieties was high enough to permit hydration of the membrane pores.

Two-Component IPNs. In order to increase the degree of polymerization while decreasing the DUP concentration, IPN membranes were prepared with 0.25M DUP and 1.5M acrylonitrile (AN). DUP/AN IPNs were first formed in polystyrene beads and found to be stable as long as the double bond concentration of the overall mixture was 2M. Membranes, however, made with this DUP/AN ratio, as well as with others having somewhat lower AN levels and correspondingly higher DUP levels, did not permit any Cu(II) transport, probably due to the DUP's lack of mobility in the predominately AN copolymer.

Three-Component IPNs. Highly stable IPN membranes were prepared from polypropylene when conclusions reported earlier with the ion exchange/coordination bifunctional polymers were extended to the membranes. Since neutral compounds with a phosphoryl oxygen moiety are important metal ion coordinators, tri(undecenyl)phosphine oxide (TUP) was synthesized and copolymerized with DUP. The less hydrophilic TUP could thus be polymerized with DUP in order to assist in the metal ion transport. It could also be expected to lower the degree of polymerization when AN is a comonomer thus allowing for high ligand mobility. A DUP/TUP/AN concentration of 0.40M, 0.28M, 0.85M, respectively, was found to offer the best compromise between stability and metal ion transport after polymerization within the polypropylene membrane: a permeability coefficient of 1.25×10^{-4} cm/sec remained constant over 36 days so that, under the conditions of the present experiment, the membrane was found to be stable over an indefinitely long period of time.

Molecular Complexation Reactions by Polymer-Supported Reagents

The complexation of neutral molecules by ligands covalently bound to polymer supports requires the study of a set of parameters different from those involving the complexation reactions of metal ions. While the latter emphasizes ion exchange mechanisms, the former relies on reactions such as acid/base interactions for selectivity. The role of the polymer support in molecular complexation reactions has been the focus of this part of our research: studies are being undertaken to determine whether the polymer acts only as an inert matrix on which to bond appropriate ligands, or whether it can also influence the ligand-molecule interaction.

Polystyryl-bound dimethylamine ligands can be expected to complex both inorganic and organic acids.

Benzoic acids, in particular, can be used to evaluate the selectivity of the ligand to variations in the electron density at the active site by determining the amount complexed from solution as electron withdrawing/donating substituents on the aromatic ring increase or decrease the acidity of the functional group. The interaction may be summarized by equation 3.

The selectivity of the ligand can be quantified across a series of substrates by determining the binding constant of each substrate via its adsorption isotherm and corresponding y-reciprocal plot (equation 4).

$$c/r = (1/S_t)c + (1/KS_t) \qquad (4)$$

In this equation, c is the milliequivalents (mequiv) of substrate in solution at equilibrium per mL solution, r is the mequiv of substrate bound per gram of polymer, S_t is the saturation capacity of the polymer in mequiv per gram, and K is the binding constant (22).

In the initial studies carried out to date, polystyrene-supported dimethylamine polymers crosslinked with 2 and 15% divinylbenzene (DVB) were complexed with m-chlorobenzoic acid and the adsorption isotherms defined. The corresponding binding constants were calculated to be 126.19 and 67.99 M^{-1}, respectively. It is thus evident that the polymer support affects the strength of the binding interaction between the ligand and substrate. Mechanistic studies to understand the basis for this influence and its use in the design of polymers for selective molecular complexation reactions are in progress.

Acknowledgment. We gratefully acknowledge the support of the Department of Energy, Office of Basic Energy Sciences, through grant DE-FG05-86ER13591.

Literature Cited

1. Cram, D.J.; Cram, J.M. Acc. Chem. Res. 1978, 11, 8.
2. Pedersen, C.J. J. Am. Chem. Soc. 1967, 89, 7017.
3. Lehn, J.M. Pure Appl. Chem. 1980, 52, 2441.
4. Timko, J.M.; Helgeson, R.C.; Newcomb, M.; Gokel, G.W.; Cram, D.J. J. Am. Chem. Soc. 1974, 96, 7097.
5. Trainor, G.L.; Breslow, R. J. Am. Chem. Soc. 1981, 103, 154.
6. Wulff, G.; Grobe-Einsler, R.; Vesper, W.; Sarhan,A. Makromol. Chem. 1977, 178, 2817.
7. Smid, J. Pure Appl. Chem. 1982, 54, 2129.
8. Dietrich, B.; Lehn, J.M.; Sauvage, J.P. Tetrahedron Lett. 1969, 34, 2885.
9. Polymer-Supported Reactions in Organic Synthesis; Hodge,P.; Sherrington,D.C., Eds., Wiley: Chichester, 1980.
10. Helfferich, F. J. Phys. Chem. 1965, 69, 1178.
11. Janauer, G.E.; Ramseyer, G.O.; Lin, J.W. Anal.Chim. Acta 1974, 73, 311.
12. Alexandratos,S.D. Sep. Purif. Methods 1988, 17, 67.
13. Alexandratos,S.D.; Wilson,D.L. Macromolecules 1986, 19, 280.
14. Alexandratos,S.D.; Crick, D.W.; Quillen, D.R. Ind. Eng. Chem. Res. 1991, 30, 772.
15. Alexandratos,S.D.; Bates, M.E. Macromolecules 1988, 21, 2905.
16. Choppin, G.R. Sep. Sci. Technol. 1981, 16, 1113.
17. Baker, R.W.; Tuttley, M.E.; Kelly, D.J.; Lonsdale, H.K. J. Membr. Sci. 1977, 2, 213.
18. Sperling, L.H. Interpenetrating Polymer Networks and Related Materials; Plenum: New York, NY, 1981.

19. Alexandratos, S.D.; Strand, M.A.; Callahan, C.M. In Advances in Interpenetrating Polymer Networks; Klempner, D.; Frisch, K.C., Eds.; Technomic Publishing: Lancaster, PA, 1990, Vol. 2; pp. 73-99.

20. Danesi, P.R.; Chiarizia, R.; Castagnola, A.J. J. Membr. Sci. 1983, 14, 161.

21. Alexandratos, S.D.; Strand, M.A. Macromolecules 1986, 19, 273.

22. Connors,K.A. Binding Constants; Wiley: New York,NY, 1987.

RECEIVED January 30, 1992

Chapter 15

Solvent Extraction Methods for the Analytical Separation of Environmental Radionuclides

W. Jack McDowell and Betty L. McDowell

East Tennessee Radiometric/Analytical Chemicals (ETRAC), Inc., 10903 Melton View Lane, Knoxville, TN 37931

In many cases solvent (liquid-liquid) extraction offers considerable advantages in the separation of radionuclides from environmental samples over methods such as ion exchange and precipitation. Although solvent extraction separations can be coupled effectively with other separation methods, they often offer special advantages themselves in that alpha, beta, and gamma radiation can all be counted directly and effectively in the organic extract. This advantage can shorten procedures dramatically. Several procedures will be outlined including methods of concentrating and counting cesium, strontium, radium, uranium, thorium, polonium and the tri-valent actinides. Methods of counting and collection of identifying spectra for beta-, gamma-, and especially alpha-emitters are included.

In 1805 Bucholtz reported that uranyl nitrate was soluble in diethyl ether(1) and an early purification of uranyl nitrate (1842) was by recrystallization from ether.(2) Solvent extraction or liquid-liquid extraction is now a familiar tool in heavy chemicals industries such as metallurgy and in analytical chemistry, but it was not widely used before the 1940's.(3) The solvent extraction of metals developed to its present sophisticated state largely because of, and certainly parallel with, the Manhattan Project. The partition of uranyl nitrate between aqueous nitric acid solutions and ether provided the means of purifying the large quantities of uranium needed for both the cyclotron and gaseous diffusion isotope separation processes. This same extraction system was also used to separate uranium from irradiated fuel elements, leaving plutonium behind.(4) The use of polyethers and high-molecular-weight ketones for such separations logically followed.(3) Some of these extractants are still used today. All of these extractants are normally used without dilution by any inert solvent and all require a high aqueous anion concentration, usually nitrate or chloride.

0097–6156/92/0509–0206$06.00/0
© 1992 American Chemical Society

Seeking other ether-like, aqueous-immiscible compounds with electron donor coordinative properties led early to the consideration of tributylphosphate (TBP). It was not used in early work because of its high viscosity, high density and difficulty in stripping, i.e., it was too good an extractant. We now know that dilution of TBP with an inert solvent (diluent) solves these problems, but this idea was quite foreign to the art and practice of solvent extraction in the 1940's.(3) All of the ethers, ketones, phosphates and similar compounds can be thought of as neutral coordinators. That is, they simply supply, in the organic phase, the coordination required by the metal that is provided by the water-oxygen in the aqueous phase. With this type of extraction, the anion for the metal in the aqueous phase must be transferred to the organic phase in the stoichiometrically required amount. Thus, the neutral coordinators are also neutral species extractants as shown in the equation below.

$$M^{+i} \cdot nH_2O + iX^- + nB \rightleftharpoons MX_i \cdot nB + nH_2O \qquad (1)$$

More recently-developed solvent extraction systems generally consist of an *extractant* dissolved in a *diluent* in contact with the aqueous phase. Neither the diluent nor the extractant should distribute to the aqueous phase and, of course, the compound formed between the extractant and the metal (ion or molecule) must also be highly organophilic. Extractants can, in general, be classified as neutral species extractants (note above) and cation exchangers. The salts of high-molecular-weight amines, are sometimes called liquid anion exchangers; however, equilibriumwise they can just as properly be thought of as neutral species extractants, and in most cases kinetic studies have not been done to determine which mechanism is correct.

Neutral-Species Extractants. Ketones, ethers, linear poly-ethers, and more recently, macrocyclic or "crown" ethers all have useful properties as extractants if they are sufficiently water insoluble. Sulfoxides, the trialkyl phosphates, the trialkylphosphine oxides, and the carbamoylmethylphosphonates are among the stronger coordinators classified as neutral species extractants. These compounds usually coordinate directly with the metal ion and supply coordination that would be supplied by water in the aqueous phase.

The acid salts of high-molecular-weight primary, secondary, and tertiary amines and quaternary ammonium compounds can also be classified as neutral species extractants. They form complexes with metals through the aqueous anion that is present. The strength of the complexes depend strongly on the ability of the aqueous-phase anion to attach to the metal cation. Bonding in the organic phase appears to be entirely through the aqueous anion. These compounds can be thought of as amine (or ammonium) salts of the anionic metal-anion complex or as adducts of the neutral metal salt with the amine salt. The general conclusion that extraction by such reagents is anion exchange or neutral-species transfer cannot be made. These organic-phase amine-salts are often stronger extractants than ethers and other oxygen donors. They have selectivity properties that depend both on the structure of the amine and the identity of the aqueous-phase anion.

It is characteristic of all of these neutral species extractants that they vary widely in their selectivity and extractive strength as the concentration of both the aqueous-phase anion and acid vary. The first because the concentration of the neutral metal species in the aqueous phase is varied and the second because the acid competes with the metal for bonding to the extractant molecule(s). Thus a large degree of control over the selectivity of the system is available.

Cation Exchange Extractants. This class of extractants includes phenols, branched alkyl carboxylic acids, alkyl phosphoric acids, diketones, and alkyl-aryl sulfonic acids. The last group listed, sulfonic acids, are analogous to sulfonic-acid cation exchange resins and have very little selectivity. Diketones, alkyl phosphoric acids and carboxylic acids can provide both cation exchange functions and coordination functions. This feature has made bis(2-ethylhexyl)phosphoric acid one of the most versatile and powerful extractants of this type.(5) The equation below illustrates simple cation exchange extraction.

$$M^{+i} + iHA \rightleftharpoons MA_i + iH^+ \qquad (2)$$

Synergistic Extractant Systems. In a synergistic system two extractants are mixed in a diluent and the extractive strength of the two extractants mixed is greater than the sum of their extractive strengths when used separately. Such synergistic systems usually consist of a cation exchange extractant and a neutral coordinative extractant. The coordinator either replaces water in the organic-phase complex making it more organophilic or adds to a coordinatively unsaturated organic-phase complex making it more stable. Typical synergistic systems are phosphoric acids plus phosphine oxides and thenoyltrifluroacetone plus a quaternary ammonium chloride or a phosphine oxide. A general equation for such a reaction is:

$$M^{+i} + iHA + nB \rightleftharpoons MA_i \cdot nB + iH^+ \qquad (3)$$

Solvent Extraction Relationships. In a solvent extraction system, the two phases are immiscible, but under proper conditions phase-transfer of one or more species can occur across the organic/aqueous interface. Expansion of the interface by gentle agitation of the vessel containing the two phases allows phase equilibrium to be attained quickly, usually in 1 to 2 minutes. In such an extraction system, the distribution of a metal, M, between the two phases at equilibrium is described by a distribution coefficient, D_M, where [M] represents the concentration of the metal.

$$D_M = \frac{[M]_{org}}{[M]_{aq}} \qquad (4)$$

This is true at any phase ratio, V_{org}/V_{aq}. However, it should be remembered that D_M is a concentration ratio, and the total amount of metal recovered depends also on the phase ratio. If we call the ratio of the total metal recovered the recovery factor, F_R, the relationship,

$$F_R = D_M \left(\frac{V_{org}}{V_{aq}} \right) \qquad (5)$$

describes the total org/aq metal ratio.

The percent of the metal transferred to the organic phase can be expressed as:

$$\%R = 100 \frac{V_{org}[M]_{org}}{V_{org}[M]_{org} + V_{aq}[M]_{aq}} \qquad (6)$$

and this can be simplified to:

$$\%R = 100 \frac{D}{D + \dfrac{V_{aq}}{V_{org}}} \qquad (7)$$

Thus, if D_M is 1000 (not an unusual value) and V_{org}/V_{aq} is 1 then F_R is also 1000, and the percent recovered is $100(1000/(1000 + 1)) = 99.9$ but if $V_{org}/V_{aq} = 1/100$ or $1/1000$ the fraction recovered is much less. For a phase ratio (org/aq) of 1:100 it is $100(1000/(1000 + 100)) = 90.9$ percent and if the phase ratio is 1:1000 the recovery will be only 50%. These simple calculations make obvious the importance of phase ratio in analytical applications of solvent extraction.

It should also be emphasized that a distribution coefficient depends on the composition of both the organic and aqueous phases. If the composition of either phase is changed, the distribution coefficient can change. It is obvious that one should be aware of distribution coefficients and phase ratios in following, and more particularly in modifying or developing, analytical procedures based on solvent extraction.

Stripping or Back-Extraction. Both of these terms are used to mean reversing the extraction equilibrium and transferring the desired element from the pregnant organic phase to an aqueous phase. This is necessary whenever the determination step is based on aqueous chemistry as for example a precipitation followed by gravimetric analysis or an aqueous-phase titration. It is also important when the extracted element is perceived as a highly hazardous substance (radioactive and/or toxic) and needs to be as completely removed from the organic phase as is possible. It has been the experience of the authors that this degree of stripping is possible in all cases with which we are familiar. For some neutral extractants water stripping is effective, for amine-salt extractants changing the aqueous anion form nitrate to chloride or perchlorate or washing the organic phase with sodium or ammonium carbonate is successful. For cation exchange systems several small scrubs with a strong acid are sufficient. Remaining amounts of radioactivity are easily reduced below a detection level of 0.001 cpm/mL.

For analytical purposes, many, if not most, radiometric determinations can be done as expeditiously from an organic phase as from an aqueous phase. Gamma counting can, of course, be done equally well in an aqueous or organic phase, and beta counting, if done by liquid scintillation methods, can be more accurate if the sample is in an organic phase. The principal reasons for this are (1) that the amount of quenching material introduced to the cocktail is minimal, (2) quenching is more constant, and (3) quenching is more under the control of the analyst. In the determination of alpha activity, electroplating can be done from organic extracts by modifying them with alcohol or other polar solvents plus acid. Some organic extracts can be dried on a hotplate or dried and flamed (phosphate extractants are the exception) followed by plate counting. Alpha counting can be done directly in the organic phase (extractive scintillator phase) using some forms of liquid scintillation, especially   Photon/Electron-Rejecting-Alpha-Liquid-Scintillation   (PERALS) spectrometry. Thus the versatility and selectivity of solvent extraction is enhanced by the ability to count directly in the organic extract.

Procedures

Sample Preparation. Samples must be placed in solution (if not already in solution) by any suitable method that is compatible with the sample matrix and the chemistry to follow.(6-9) With the sample in solution, a further concentration or preliminary separation may be necessary. For example: (1) an evaporation to concentrate the sample, (2) a precipitation or coprecipitation to concentrate and separate the desired nuclide from large amounts of other ions, or (3) a solvent extraction or ion exchange step for the same purpose.

As an example, let us consider a soil sample containing organic material. First, the organic material must be oxidized. One- to two-gram portions of the sample are weighed into crucibles and ignited at 600°C until all the organic material is destroyed. The material is then placed in solution by some standard method.(6-10) The final step of sample preparation must always include placing the sample in the anion matrix and pH appropriate for the desired separation or determination.

The procedures outlined below are not intended to be a comprehensive list or review of solvent extraction analytical separations for radionuclides but, rather, to provide a condensed guide to some recently developed methods (primarily at Oak Ridge National Laboratory, Oak Ridge, Tennessee) that illustrate the advantages offered by this approach to radiometric analyses. Selected procedures follow.

Uranium Activity on Cellulose Filters.(10) Cellulose air filters can be easily placed in solution and assayed as follows: The filter paper (up to 2-in diam.) is placed in a 2-dram, screw-cap borosilicate glass vial and heated (open, without cap) in an oven or furnace at 500°C for 2 hours. After removing and cooling, add 2 to 5 drops of concentrated nitric acid, 3 drops of 30% hydrogen peroxide, and 1 drop of a saturated aluminum sulfate solution to the vial. Next, heat the vial to 200°C to remove the nitric acid. The solids remaining are redissolved in a solution 1.0 M in Na_2SO_4 and 0.01 M in H_2SO_4. A measured quantity of 0.1 M high-molecular-weight tertiary amine sulfate in toluene is then added to the vial and the two phases are equilibrated for 2 to 3 min. After the phases have separated, the organic phase can be dried or electroplated (after addition of dimethylsulfoxide or alcohol and hydrochloric acid) on a planchet and counted. If the tertiary-amine-sulfate extractant also contains a fluor and an energy transfer agent such as naphthalene, the organic extract can be counted directly on a PERALS spectrometer. Counting efficiency by this method is virtually 100%, alpha spectra thus obtained have beta/gamma backgrounds comparable to those in plate counting, and spectra usually show sufficient energy resolution to allow identification of the isotopes of uranium present (see Figure 1).

Gross Alpha in Environmental Materials.(*11*)   With the sample in solution in nitric acid, add 0.5 g of $LiClO_4$ and 1.0 to 1.5 mL of 0.1 M $HClO_4$. Evaporate under heat lamps or in an aluminum block at 160°C until boiling of the sample ceases and the first perchloric acid fumes appear. Nitric acid + 30% hydrogen peroxide can be used to clarify the sample instead of perchloric acid if laboratory rules forbid perchloric acid. However, perchloric acid is safe and more effective. Cool the beaker and add 5 to 7 mL of water to dissolve the viscous (or solid) residue. Measure the pH of the solution; it must be between 2 and 3.5. If it is not, carefully neutralize the excess acid with sodium hydroxide or add nitric acid to increase acidity. Transfer the solution to a small equilibration vessel and add a measured quantity of O.1 to 0.2 M bis(2-ethylhexyl)phosphoric acid in toluene. Equilibration will transfer most of the alpha-emitting nuclides, with the exception of radium and radon, to the organic phase. Lanthanides are also extracted. For plate counting methods, the activity must be stripped into an aqueous phase using strong nitric acid or ammonium carbonate and electroplated or dried. Direct counting of the organic phase on a PERALS spectrometer may be done if the appropriate extractive scintillator is used in the extraction step. Identification of the nuclides as well as an accurate quantification of them is usually possible with this method.

Uranium and Thorium. (*W.J. McDowell and G.N. Case, unpublished results.*) With the sample in solution, add sulfuric acid and sodium sulfate and convert to a sulfate system at pH 1 to 2. An initial extraction into an extractant composed of a tertiary-amine-sulfate 0.1 to 0.2 M in toluene will quantitatively remove uranium leaving thorium behind. A second extraction from the same aqueous using an extractant containing a branched primary amine sulfate will remove thorium. The extractions must be in this order since the primary amine will also extract uranium. Each organic extractant phase is then sampled and counted as desired. Again, direct counting of the organic phase can be accomplished if the extractant is also a scintillator. Spectra of normal uranium in secular equilibrium will show the double peaks due to ^{238}U and ^{234}U (Figure 1). However, natural processes can concentrate ^{234}U and artificial processes have concentrated $^{234-235}U$; thus, not all uranium now in the environment is "normal" uranium. Spectra of uranium enriched in ^{235}U will show primarily ^{234}U, and spectra of tailings from enrichment processes will show a predominance of ^{238}U.

In many non-phosphate samples, uranium can be coprecipitated with magnesium hydroxide or otherwise concentrated and extracted from a sulfate system into an extractive scintillator containing a high-molecular-weight tertiary amine sulfate.

Uranium and Thorium in Phosphates.(*12*)   These elements can be separated from a variety of phosphate-containing materials, e.g., fertilizers, bones, teeth, animal tissues, and wastes using this procedure. The sample is oxidized (if necessary), dissolved and placed in a nitrate or nitrate-perchlorate solution. To this solution, sufficient aluminum nitrate is added to complex the phosphate. This solution is then

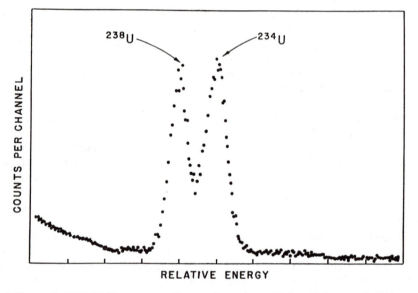

Figure 1. PERALS spectrum of normal uranium. (Adapted from ref. 12.)

contacted with a toluene solution of trioctylphosphine oxide (TOPO). Both uranium and thorium are transferred to the organic phase. The organic phase is then stripped with an equal volume of 0.5 M ammonium carbonate solution, the ammonium carbonate is evaporated, the sample is converted to nitrate or nitrate/perchlorate, and any entrained organics are destroyed with nitric acid and hydrogen peroxide or with nitric acid and perchloric acids. The clear solution is then converted to a sulfate system and treated as in the previous procedure.

Polonium:(13) The radioisotopes of polonium (usually ^{210}Po) have been difficult to analyze with accuracy using the conventional methods. The procedure outlined here is, however, simple, rapid, and accurate. With the sample in solution, add 3 to 5 mL of concentrated phosphoric acid and evaporate to remove other acids. Transfer this phosphoric acid solution to a small equilibration vessel using 3 to 5 mL of water. Add 1 mL of 0.1 M HCl. Add a measured volume, 1.2 to 1.5 mL, of a solution of TOPO, 0.1 to 0.2 M, in toluene and equilibrate. This is a highly selective separation of polonium from other radionuclides with the possible exception of the beta/gamma emitting bismuths. Quantitative stripping and transfer of the polonium to a plate is difficult but the use of an extractive scintillator and counting on a PERALS spectrometer is rapid and simple and the results are quite accurate. Because of the minimal chemical manipulations required, the accuracy of this determination can easily be better than $\pm$ 1%.

Plutonium.(14) Plutonium can be chemically separated from all other elements except neptunium and counted quantitatively by this procedure, and the 4.78 MeV ^{237}Np peak can be resolved by its energy difference from 5.15 MeV $^{239\text{-}240}$Pu. The sample should be in solution in a 10 mL volume of 3 to 4 M total NO_3^- and 0.5 to 1.0 M HNO_3. The plutonium is then quantitatively reduced to Pu(III) with ferrous sulfate (approx. 50 mg) and reoxidized to Pu(IV) with a similar quantity of sodium nitrite. This solution is contacted with not less than 1/4 its volume of 0.3 M trioctylamine nitrate solution in toluene. The distribution coefficient for plutonium is about 4000, that for uranium is 1.0, for iron is 0.01 and for other elements is much lower. Equilibrate and separate the aqueous phase. Wash the organic phase with two 1/4-volume portions of 0.7 M HNO_3 to remove any remaining uranium. The aqueous phase from the first equilibration and the washes can be combined and analyzed for uranium, if desired. Plutonium can be stripped from the organic phase either with perchloric acid (perchloric acid is strongly extracted but amine perchlorates are not extractants) or, after doubling the organic phase volume with 2-ethylhexanol, with 1 N H_2SO_4. Drying or electroplating for plate counting can be done either directly from the organic extract or after stripping. Alternatively, the plutonium can be reextracted into an HDEHP-containing scintillator. In the first instance, when the aqueous phase is perchlorate, and at a pH of 2 to 3, an HDEHP-containing scintillator (ETRAC's *ALPHAEX*) can be used. When the plutonium is stripped into a sulfate system, a scintillator containing a branched primary amine sulfate (ETRAC's *THOREX*) is required. Either extract can be counted directly on the **PERALS** spectrometer.

Radium.(*15*) Radium can be separated from other alkaline earth elements using a synergistic extractant consisting of 0.1 *M* high-vacuum-distilled, high-molecular-weight neocarboxylic acid and 0.05 *M* dicyclohexano-21-crown-7. Extraction is from a basic solution. The following procedure allows separation from many (if not all) other cations. The sample solution is first spiked with [133]Ba or [224]Ra, and 10 mg of barium carrier is added. Radium is then precipitated from the sample solution as barium/radium sulfate. After separation and washing, the precipitate is converted to the carbonate by heating with a saturated potassium or ammonium carbonate solution. The separated barium/radium carbonate is washed with a minimum of water or dilute ammonium carbonate solution and the solids dissolved in dilute acid, the pH of this solution is adjusted to between 9 and 10, and the radium is extracted selectively into the above extractant. The extraction is quantitative, and any chemical losses in the coprecipitation and metathesis steps can be corrected by gamma counting the [133]Ba or alpha counting of [224]Ra. The extract can be dried or plated. Radium can also be counted on a PERALS spectrometer, initially showing a single alpha peak each for both [226]Ra and [224]Ra. All the alpha peaks from the daughters of both radiums can be seen if one waits for their ingrowth (See Figure 2). The percent relative error in this determination using the **PERALS** spectrometer, with appropriate counting statistics, is usually less than $\pm$ 5% and can be much less.

Strontium.(*16*) This procedure has not been examined in sufficient detail to define all the conditions categorically but the work that has been done indicates sufficient confidence in and advantage to the method warrant it's inclusion here.

    The sample must be in an acid nitrate solution. Dissolution of solids will, of course, be done in the manner most suited for the sample. However, solution in a pressurized vessel with a mixture of nitric, hydrochloric, and hydrofluoric acids(*7*) is suggested for mineral or soil samples.

    If the sample has large amounts of other alkali, alkaline earth or transition metal ions present, a preliminary separation is recommended. Adjust the acidity to 0.2 *M* HNO_3 and contact the aqueous phase with 1/4th its volume of an organic phase composed of didodecylnaphthalene sulfonic acid (HDDNS) and dicyclohexano 18-crown-6 (DC18C6) in toluene in a mole ratio of 2 to 1. At 0.1 *M* HDDNS and 0.05 *M* DC18C6, the distribution coefficient for strontium, D_{Sr} = 6000, that for calcium is 80, potassium is 46, rubidium is 16, and other ions are lower except for barium which is also 6000. Using lower concentrations of the two extractants, HDDNS and DC18C6, will reduce all the extraction coefficients approximately in direct proportion to the concentration and may result in a better separation since this reagent dilution will place the D_M values of many unwanted ions at less than unity. The organic phase will now contain all of the strontium and barium and only small amounts of other ions. The phases should be separated and the metal ions in the organic phase stripped into a new aqueous phase. This is necessary because the HDDNS-containing extract cannot be added to a scintillator or be part of a scintillator because HDDNS is brown colored and would severely quench the beta emitting ^{90}Sr/[90]Y. Freshly purified HDDNS is not colored but becomes so within a matter of hours. HDDNS is a very strong acid and stripping is difficult. However, two methods of stripping are known to work: 1) Several small volume equilibrations with

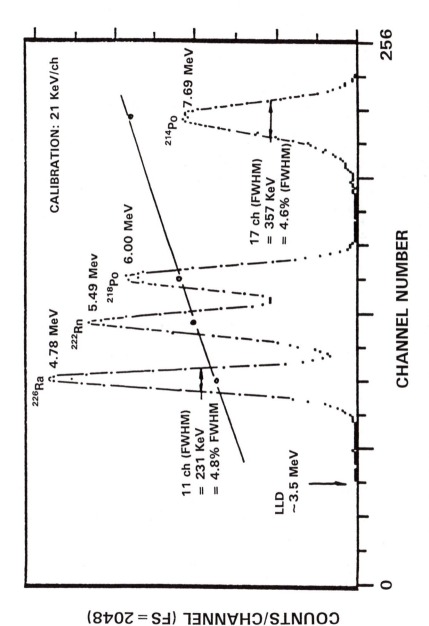

Figure 2. PERALS spectrum of radium-226 and daughters. (Adapted from ref. 15.)

6 to 8 M HNO_3 or   2) A single equilibrations with a 0.2 M aqueous solution of pentasodiumdiethylenetriaminepentaacetate (Na_5DTPA or Versenex-80).   In either case the aqueous phase strip solution must be boiled with nitric acid and 30% H_2O_2 to destroy Na_5DTPA and/or color in the strip solution.

Beginning with the aqueous phase from the steps above or from an aqueous phase containing only small amounts of other ions (for example water samples), the required steps are: 1) Adjust the aqueous phase to a pH > 9 and extract the strontium into a toluene solution containing 0.1 M high-vacuum-distilled, high-molecular-weight neocarboxylic acid in the sodium-salt-form and 0.05 M DC18C6.   This extraction will further separate strontium from other metal ions. Work to date does not allow a statement of all relevant separation factors, but the combination of the two steps is expected to give good separation of strontium over cesium, potassium and most other ions.   Strontium is selected over barium in this step by a factor of at least 2.   Selectivity for strontium is better in the latter extraction than in the first.

If the last extractant is also a scintillator such as ETRAC's STRONEX, the equilibrated organic phase can be counted directly in a beta-liquid-scintillation counter or in a PERALS spectrometer.   The carboxylic acid is not colored and does not quench.   The PERALS spectrometer provides better beta-energy resolution and has only slightly lower counting efficiency for betas and thus may offer some advantage if both ^{90}Sr and ^{89}Sr are required in the same sample. The PERALS spectrometer will provide better separation of the 0.546 MeV ^{90}Sr, the 1.48 MeV ^{89}Sr, and the 2.28 MeV ^{90}Y.   In addition, if radium is present, the pulse shape discrimination feature of the PERALS spectrometer can be used to reject the contribution from radium alphas.

The assay for ^{89}Sr and ^{90}Sr together will require that standards be used to set up counting efficiencies in two regions of interest. Backgrounds for these regions of interest would also be required and the contribution of the ^{89}Sr (1.49 MeV beta) to the ^{90}Sr (0.546 MeV beta) count will be required. With this information, a single count of a mixture would yield information on the amount of both isotopes, the accuracy depending on the total count and the relative quantities of the two nuclides.

An alternate scheme could be set up to determine ^{90}Sr from the increase in count due to the ingrowth of ^{90}Y.

Cesium.(*W.J. McDowell, Manuscript in preparation*)   A synergistic mixture of purified HDDNS acid and di(tert.-butylbenzo)21-crown-7, 0.1 and 0.05 M, respectively, in toluene extracts cesium strongly and selectively from dilute nitric acid solutions. Unfortunately, as of this writing, the crown ether is not commercially available. It is hoped that in the future it will be. An alternative extractant is a high-molecular-weight substituted phenol.(*17*) These organic extracts can, of course, be gamma-counted directly in the organic phase.

Acknowledgments

Much of the work reported in this paper was performed at Oak Ridge National Laboratory under the sponsorship of the Department of Energy, Basic Energy Sciences Division. G.N. Case was the person responsible for most of the laboratory work.   Oak Ridge Detector Laboratory (ORDELA) has provided space for the laboratory of ETRAC, Inc. where the most recent work was done.

Literature Cited

1. Bucholtz, C.F. *J.Chem. von A.F. Gahlen*, **1805**, 4(17), 134.
2. Peligot, E. *Ann. Chem. Phys., 3rd Series*, **1842**, 5, 5-47, 102.
3. Coleman, C.F.; Leuze, R.E. *J. TN Acad. Sci.* **1978**, 53(3),pp.
4. Seaborg, G.T. *U.S. ERDA Rept. PUB-112*, **1978**, Berkeley, CA.
5. Shoun, R.R.; McDowell, W.J. In *Actinide Separations;* Navratill, J.D.; Schulz, W.W., Ed.; ACS Symposium Series, No. 117; American Chemical Society: Columbus, OH, 1980.
6. Farrell, R.F.; Mathes, S.A.; Mackie, A.J. *A Simple Low-Cost Method for the Dissolution of Metal and Mineral Samples in Plastic Pressure Vessels;* Bureau of Mines Report of Investigations; No. 8480, **1980**.
7. Matthes, S.A.; Farrell, R.F.; Mackie, A.J. *A Microwave System for the Acid Dissolution of Metal and Mineral Samples;* Bureau of Mines Technical Progress Report-120, **1983**.
8. Sill, C.W. *Anal. Chem.*, **1969**, 41, 1624.
9. Sulcek, Z.; Povondra, P. *Methods of Decomposition in Inorganic Analysis*; CRC Press: Boca Raton, FL, 1989.
10. McDowell, W.J.; Case, G.N. *A Procedure for the Determination of Uranium on Cellulose Air-Sampling Filters by Photon-Electron-Rejecting Alpha Liquid Scintillation Spectrometry*; ORNL/TM 10175; Aug 1986.
11. McDowell, W.J. *Alpha Counting and Spectrometry Using Liquid Scintillation Methods*; Radiochemistry Series; NAS-NS-3116: Technical Information Center, U.S. Dept. of Energy: Oak Ridge, TN, 1986; 88-89.
12. Bouwer, E.J.; McKlveen, J.W.; McDowell, W.J. *Health Physics* **1978**, 34, 345.
13. Case, G.N.; McDowell, W.J. *Talanta* **1982**, 29, 845-848.
14. McDowell, W.J.; Farrar, D.T.; Billings, M.R. *Talanta* **1974**, 21, 1231.
15. Case, G.N.; McDowell, W.J. *Radioactivity and Radiochemistry* **Fall 1990**, 1(4), 58-69.
16. Bryan, S.A.; McDowell, W.J.; Moyer, B.A.; Baes, C.F., Jr.; Case, G.N. *Solvent Extn. and Ion Exch.* **1987**, 5(4), 717-738.
17. Zingaro, R.A.; Coleman, C.F. *J. Inorg. Nucl. Chem.* **1967**, 20, 1287-1300.

RECEIVED January 24, 1992

Chapter 16

Limiting Activity Coefficients
of Nonelectrolytes in Aqueous Solutions

D. L. Bergmann and C. A. Eckert[1]

School of Chemical Engineering, Georgia Institute of Technology,
Atlanta, GA 30332-0100

Infinite dilution activity coefficients of non-electrolytes in water are very useful for the economical design of both separation systems for specialty chemicals and for environmental control. However, the unique properties of water result in very large nonidealities and present special challenges in measurement, correlation, and prediction. A number of measurement techniques that exist specifically for determining limiting activity coefficients are reviewed. They provide a modest base of reasonably reliable data. Current estimation techniques are limited. Certainly more good data are needed for both design and correlation development.

Almost every process in the chemical industry revolves around separations of materials, and the majority of the capital cost and as much as 90% of the energy costs go toward separations. Most separation processes are multistage processes and require an accurate representation of the phase behavior involved for both process feasibility and economics.

Dilute Aqueous Solutions

In today's market, with the rising costs of imported feed stocks, high purity, high value, low volume specialty chemicals such as pharmaceuticals are becoming increasingly important. Another area of burgeoning interest is in environmental separations. Both of these trends involve separations of dilute solutions. For specialty chemicals, high purity is often necessary, requiring the removal of dilute contaminants. In some cases, as in the biochemical industry, the dilute species may be the product of interest. Environmental separations such as treatment of hazardous wastes and pollution control often involve the concentration and removal of dilute contaminants. Thus we need a better understanding of the phase behavior of these dilute solutions to facilitate design of such separations.

Aqueous solutions are of special interest for several reasons. Because of its unique properties and its great availability, water is the most commonly used solvent in industry. Water is a complex solvent which hydrogen bonds extensively; the actual

[1]Corresponding author

0097–6156/92/0509–0218$06.00/0
© 1992 American Chemical Society

structure of liquid water is not well understood (*1*). Due to its complexity, many aqueous solutions of non-electrolytes exhibit highly nonideal behavior, often resulting in regions of immiscibility. Water is also important in the new trends of the chemical industry. Most biochemical recovery processes are fed with a dilute stream of products which is almost exclusively water (*2*). Treatment of agricultural run-off waters and purification of drinking water are examples of environmental separations. One could hardly deal with the problem of cleaning up the environment without involving aqueous solutions. Thus, the ability to represent phase behavior of aqueous solutions is imperative.

Limiting Activity Coefficients

Activity coefficients are a means of representing the phase behavior of a system containing a liquid phase. For example, consider a liquid phase in equilibrium with a vapor phase. The equilibrium equation for component 1 in this system is given by:

$$x_1 \gamma_1 \phi_1^{sat} P_1^{sat} \exp\left[\frac{v_1\left(P - P_1^{sat}\right)}{RT} \right] = y_1 \phi_1 P$$

(1)

In this expression x and y are the mole fractions of component 1 in the liquid and vapor phase respectively. P is the system total pressure and P_1^{sat} is the vapor pressure of component 1. ϕ_1 is the vapor phase fugacity coefficient representing the nonideality in the vapor phase. It can be calculated from an appropriate equation of state, and at low pressures has a value close to unity. The terms ϕ_1^{sat}, the fugacity coefficient at the saturation pressure, and the exponential term, known as the Poynting correction, are small corrections that are often negligible. v_1 is the molar volume of liquid 1, R is the gas constant and T the temperature. γ_1 is the activity coefficient of component 1 in the liquid phase, representing the nonideality of the liquid phase, and has a value of unity for pure liquid 1. In typical aqueous solutions of interest, the vapor phase nonidealities are negligible, but the liquid γ often deviates from unity by orders of magnitude. If one has a way of determining the value of this activity coefficient the phase equilibrium of the system can be correlated using equation 1.

The activity coefficient of a component in a mixture is a function of the temperature and the concentration of that component in the mixture. When the concentration of the component approaches zero, its activity coefficient approaches the limiting activity coefficient of that component in the mixture, or the activity coefficient at infinite dilution, γ^∞. The limiting activity coefficient is useful for several reasons. It is a strictly dilute solution property and can be used directly in equation 1 to determine the equilibrium compositions of dilute mixtures. Thus, there is no reason to extrapolate equilibrium data at mid-range concentrations to infinite dilution, a process which may introduce enormous errors. Limiting activity coefficients can also be used to obtain parameters for excess Gibbs energy expressions and thus be used to predict phase behavior over the entire composition range. This technique has been shown to be quite accurate in prediction of vapor-liquid equilibrium of both binary and multicomponent mixtures (*3*).

The limiting activity coefficient is also of great theoretical interest. At infinite dilution each solute molecule is surrounded by only solvent molecules, and the most nonideal conditions are represented. γ_1^∞ is in fact an excess property, so like-pair interactions are normalized out. Since only unlike-pair interactions are involved, any composition dependence of the solute on the properties of the mixture are removed, thus allowing the unlike-pair interactions to be studied directly. An infinitely dilute solution is also much simpler to model when using a statistical mechanical approach. The investigation of aqueous systems is also of theoretical importance. Due to its hydrogen bonding characteristics, water often exhibits large nonideal behavior with other compounds. Understanding such extreme behavior will help in understanding less complex mixtures. Thus limiting activity coefficients are useful in understanding and representing dilute solution equilibrium behavior.

Techniques For Measuring γ^∞ For Aqueous Solutions

Activity coefficients at infinite dilution can be measured by several techniques, each with its advantages and disadvantages. The most established techniques are dynamic or inverse gas chromatography and differential ebulliometry. Other methods include gas stripping, headspace chromatography, and liquid-liquid chromatography. Since many systems exhibit miscibility gaps in water solutions, solubility data have also been used to obtain γ^∞. These techniques have been used by several researchers to investigate a variety of binary water systems.

The use of dynamic gas chromatography to make thermodynamic measurements was first proposed by Martin (4). The value of γ^∞ is found by measuring the retention time of a solute injected into an inert carrier gas stream passing through a gas chromatographic column containing the solvent as the stationary phase. Details of the apparatus and procedure used in this technique are given by Pecsar and Martin (5) and Shaffer and Daubert (6). Equipment for such measurements can be found commercially, and the technique is well established. The method is quick and simple. The solute need not be of extreme purity and the detector need not be calibrated, since only retention time data are required. The technique is best suited for the study of highly volatile solutes in solvents of low volatility; it is inapplicable for systems where the solute has low relative volatility. The technique has been extended to systems with moderately volatile solvents by presaturating the carrier gas with the solvent (7), and correcting for solvent stripping from the column. The main source of error in this technique is adsorption of the solute at various interfaces within the column. The study of polar solutes in nonpolar solvents is hindered by adsorption on exposed solid support material (8-9). For solutes that are appreciably immiscible in a solvent, as in many aqueous systems, adsorption can occur on the vapor-liquid interface, and overloading the column where peak width, shape and position change may also become a problem. Several researchers have used gas chromatography to measure γ^∞ for solutes such as chlorohydrocarbons, oxygenated compounds, aromatics and hydrocarbons in water (5-7,10-16).

Another technique for γ^∞ is differential ebulliometry, first proposed by Gautreaux and Coates (17). In this method γ^∞ is found by measuring the change in the boiling point of a solvent with the addition of a dilute amount of solute under constant pressure. Descriptions of the apparatus and procedure involved are given by Scott (18) and Trampe and Eckert (19). The differential measurement negates the effects of pressure fluctuations that affect the temperature measurements and precludes the need for thermometer calibration. The technique has been tested extensively and found to be very accurate. Several solutes can be run simultaneously in the same solvent; however, it takes several hours to obtain a value of γ^∞. Since the experiment requires a heat sink, measurements are mostly made at higher than room temperatures. The technique is limited to measuring systems with relative volatilities between approximately 0.05 and 20. This upper volatility constraint has limited researchers to studying only alcohols and ketones in water. It has been more widely used to measure γ^∞ for systems in which water is the dilute component (20-24).

Headspace chromatography measures γ^∞ by using gas chromatography as an analytical tool to measure the composition of a low-volume vapor phase in equilibrium with a dilute liquid solution. Great care is required in proper sampling and calibration of the detector. The technique can be used with nonvolatile or volatile solvents; the solute must be volatile enough to be measurable in the vapor phase. For solutes that are appreciably immiscible in the solvent, knowing the accurate composition of the liquid phase becomes the limitation. Once the detector has been calibrated accurately, the measurement is quick, and a series of solutes can be measured in the same solvent simultaneously. A detailed description of the apparatus and procedure involved is given by Hassam and Carr (*25*). This technique has only recently been explored as a way of measuring γ^∞ for a variety of solutes in water (*26-29*).

Gas stripping (*30*), sometimes called the dilutor method, involves bubbling an inert gas through a dilute solution to strip the solute. The limiting activity coefficient is related to the decay in concentration of the solute in the exit gas stream, normally measured by gas chromatography. Since a decay rate is measured and not a concentration, the gas chromatograph need not be calibrated, but a constant calibration factor must be assumed. A detailed description of the equilibrium cell and procedure used in this measurement is given by Richon et al (*31*) and Richon and Renon (*32*). The technique is applicable only for volatile solutes in low volatile solvents, and the solute must be volatile enough to be stripped in a reasonable amount of time. When the relative volatility between the solute and solvent becomes very large, which is typical for systems with large γ^∞ values, it becomes difficult to obtain equilibrium in the column. Limiting activity coefficients for some alcohols, ketones, acetates and benzene have been measured in water (*33-37*).

Liquid-liquid chromatography (LLC) is a technique that can also be used to obtain ratios of γ^∞ values. The theory of LLC as a method to probe the thermodynamics of liquid solutions has been known for years (*38*). LLC takes advantage of the fact that a solute will distribute itself between mobile and stationary liquid phases in a column, leading to a specific retention time of the solute in the column. This retention time is related to the ratio of γ^∞ of the solute in the mobile phase to γ^∞ of the solute in the stationary phase. Thus, to obtain γ^∞ of the solute in one phase requires a priori knowledge of γ^∞ in the other phase. The mobile and stationary phases must be immiscible. If they are not, the ratio of γ^∞ values measured is not that of γ^∞ in the pure solvents but the ratio of γ^∞ values in the saturated mixtures which can be appreciably different. There are several organic liquids which are almost completely immiscible in water for which there are γ^∞ data available through gas chromatography. The solute must be somewhat soluble in both liquid phases in order to be measurable (*39*). However, the main problem with this technique is avoiding the problems of adsorption on the liquid and solid interfaces in the column. It has been shown (Carr, P. W.,University of Minnesota, personal communication, 1990) that when the mobile phase has a large concentration of water, the history of the column greatly affects the retention of solutes in the column. Janini and Qaddora (*39*) and Djerki and Laub (*40*) have used this technique to measure γ^∞ for various oxygenated hydrocarbon solutes in water. They used OV-1 polydimethylsiloxane (PDMS) and squalane as the stationary phase respectively. Comparison of data for similar solutes in water measured by these researchers yields a difference ranging from 20-100%.

The oldest technique for obtaining values of γ^∞ for immiscible substances in water is actually an estimation relating γ^∞ of the component in water to the inverse of its solubility in water. For example, consider two liquid phases in equilibrium with each other. The equilibrium expression for component 1 in this system is given by:

$$x_1 \gamma_1 = x_1^{/} \gamma_1^{/}$$

(2)

In this expression x_1 and $x_1^{/}$ are the compositions of component 1 in the two phases. For most binary aqueous systems, when x_1, the mole fractions of component 1 in water, is small, $x_1^{/}$, and therefore $\gamma_1^{/}$ are both close to one. This results in the approximation $\gamma_1 = 1/x_1$. This approximation is valid at the solubility concentration x_1.

The question remains, however, of whether the solution is in fact infinitely dilute at a solute concentration of x_1. Only if this is true is it valid to assume that $\gamma_1 = \gamma^\infty$. Literature values of solubility data for several compounds in water were used to obtain parameters for the UNIQUAC and NRTL excess Gibbs energy expressions, and γ^∞ values for these compounds were calculated. The calculated values are compared with inverse solubility data in Table I. The inverse solubility predicts lower values of γ^∞ in all cases. However, the difference becomes smaller as the solubility decreases, and for compounds with solubility less than 0.5% the difference is less than 10%. It has been shown that these excess Gibbs energy expressions, while very useful, are not the exact representation of the composition dependence of activity coefficient; all expressions have difficulty in representing liquid-liquid equilibria (43-44). Thus, extrapolating these expressions to infinite dilution may be in error. It is therefore inconclusive as to the correctness of using the inverse solubility to calculate γ^∞.

Direct measurement of γ^∞ would confirm whether or not the solution is infinitely dilute at saturation. Lobien and Prausnitz (23) have attempted to measure this effect in a few systems by comparing the solubility limit with measurements of γ^∞ from differential ebulliometry. The systems they studied all had solubilities of a few percent, and for these systems they found significant deviations from $\gamma_1^\infty = 1/x_1$. It would be useful to have measurements for more dilute solubilities, but in this case the limiting activity coefficient becomes very large, and ebulliometry is inapplicable for high relative volatilities. Perhaps such data could be taken by ebulliometry for systems where the solute is much less volatile than water, or by chromatographic methods.

Limiting activity coefficient data for a few water solvent systems measured by various researches using the techniques discussed are shown in Table II. The γ^∞

Table I. Comparison of Inverse Solubility to γ^∞ Calculated by the UNIQUAC and NRTL Expressions with Parameters Found from Mutual Solubility Data at 25 °C

Compound	Solubility in Water[41], mole fraction	(solubility)$^{-1}$ in water	γ^∞ UNIQUAC	γ^∞ NRTL
2-Butanone	0.0763	13.1	25	21
Propionitrile	0.0358	28	38.5	37
Nitromethane	0.0355[42]	28.1	38.6	37.2
Ethyl Acetate	0.016	62.5	76	72
Methylene Chloride	0.00417	240	260	257
Benzene	0.000416	2400	2438	2440
Toluene	0.000106	9430	9470	9480
CCl_4	0.000092	10870	10930	10930

Table II. γ^∞ For Solutes in Water Measured by Several Researchers Using Various Techniques

Solute in Water	Temp°C	γ^∞ Measured By Various Techniques					
		GC[a]	Ebul.	HS	GS	LLC	(sol)$^{-1}$ [41]
N,N-Dimethylformamide	25			0.62[29]			
	50		0.89[24]				
	60		1.35[24]				
	70		2.67[24]				
Ethanol	20	6.51[5]					
	24.3	4.74[6]					
	25	3.91[14] 3.92[15]		3.76[26] 3.69[29]	3.27[35] 3.55[36]		
Acetone	25	7.56[15]		7.00[29]		21.1[40] 62[39]	
	30				7.69[36] 7.70[37]	59[39]	
Isopropanol	25	8.13[15] 8.14[14]		7.47[29]			
	45		8.8[24]				
	55		9.6[24]		12.25[34]		
	65		9.5[24]				

Continued on next page

Table II. Continued.

Solute in Water	Temp°C	GC[a]	Ebul.	HS	GS	LLC	(sol)⁻¹ [41]
				γ^∞ Measured By Various Techniques			
Isopropanol (continued)	76		11[24]				
	80		13.62[20]				
	85		11.6[24]				
	90		13.68[20]				
	100		14.00[20]				
2-Butanone	25	27.8[15]		26[26] 25.6[29]		41.2[40] 66[39]	13.1
	30				29.5[37]	72[39]	
	20			41.4[28]			
	25	50.5[14] 51.6[15]		49.2[29]	45.1[35]	206[40]	52
	40	49.5[11]					
	60	59.3					
1-Butanol	70		59.3[21] 67.8[23]				
	80		46.5[23] 57.2[21]				
	90		55.5[21]				

Solute	T					
1-Pentanol	25	197[15] 197.5[14]	194[29]	192[35]	338[40]	270
Chloroform	20	571[5] 821[16] 1000[7]				810
	20	2500[7]				
Benzene	25	2289[15] 2400[12]		1700[33] 2200[33]		2400
CCl$_4$	20	2870[5] 6300[7]				11000
Hexane	20	2940[5]				
	25	40000[12]				340000
	30	2225[5]				
	40	1465[5]				

[a] GC- Gas Chromatography, Ebul.- Differential Ebulliometry, HS- Headspace Chromatography, GS- Gas Stripping, LLC- Liquid-liquid Chromatography, (sol)$^{-1}$- Inverse mutual solubility data in water.

values range from less then one to several thousand. The system N,N-dimethylformamide in water has a low relative volatility at low temperatures. It is an example of a system in which gas chromatography and gas stripping are not applicable. Different researchers using the gas chromatographic technique have obtained very good agreement for the ethanol, isopropanol, 1-pentanol and benzene systems. There is much less agreement in the chlorinated systems. For the systems where γ^∞ is very large, ie. CCl_4 and hexane, the measurement is done exclusively by gas chromatography. However, the measured values of γ^∞ are considerably less then the inverse solubility. This disagreement is possibly due to vapor-liquid interfacial adsorption. In systems where γ^∞ was measured at the same temperature by several different techniques, there is good agreement between the gas chromatographic, headspace chromatographic and gas stripping techniques. There is, however, poor agreement with the liquid-liquid chromatographic values. This is probably due to the adsorption problems encountered with the LLC technique.

A modest data base for aqueous systems has been obtained by the use of these techniques. The data are reasonably reliable for systems with γ^∞ values less then a couple thousand and not measured by the liquid-liquid chromatography technique. A reliable data base is required in the development of predictive techniques for γ^∞. Several predictive techniques are currently available; the MOSCED (45) model has not yet been extended to aqueous systems. UNIFAC (46-48), which is really an outgrowth of ASOG (21,49) does include water, but with mixed results at best. Linear solvation energy relationships (LSER's) have been used to correlate ratios of γ^∞ values for aqueous systems (50) and may be capable of some prediction. Nonetheless, a more extensive and accurate data base is what is really needed for correlation development.

Conclusions

Limiting activity coefficients for aqueous solutions are needed for obtaining phase equilibria necessary for accurate design of separation process, especially dilute solutions involved in specialty chemical production and environmental control. Several techniques are available for the direct measurement of γ^∞. With these methods, a variety of water systems have been measured, supplying a modest database of reasonable reliability. Several systems, however, are still unreachable by these existing techniques. Systems with large values of γ^∞ are still a problem, but the inverse solubility is often the closest estimation. It would be beneficial to have a technique that is capable of measuring these systems directly. Systems with very low relative volatilities are also unreachable. New techniques would be beneficial in reaching these systems and also in the comparison and calibration of data obtained from the existing methods. Data are also necessary for the development of predictive techniques for such complex systems.

Acknowledgments

The authors gratefully acknowledge the financial support of the DuPont Company and the Amoco Chemical Company, and the advice and assistance of Leon Scott of DuPont and of Peter Carr of the University of Minnesota.

Literature Cited

(1) Symons, M. C. R. *Chem. in Britain.* **1989**, *25*, 491.
(2) Belter, P. A.; Cussler, E. L.; Hu, W. *Bioseparations*; John Wiley & Sons, Inc.: New York, NY, 1988.
(3) Schreiber, L. B.; Eckert, C. A. *Ind. Eng. Chem., Process Des. Dev.* **1971**,*10*, 572.
(4) Martin, A. J. P. *Analyst.* **1956**, *81*, 52.
(5) Pecsar, R. E.; Martin, J. J. *Anal. Chem.* **1966**, *38*, 1661.
(6) Shaffer, D. L.; Daubert, T. E. *Anal. Chem.* **1969**, *41*, 1585.
(7) Thomas, E. R.; Newman, B. A.; Long, T. C.; Wood, D. A.; Eckert, C. A. *J. Chem. Eng. Data* **1982**, *27*, 399.
(8) Parcher, J. F.; Hung, L. B. *J. Chrom.* **1987**, *399*, 75.
(9) Conder, J. R. *Anal. Chem.* **1976**, *48*, 917.
(10) Hardy, C. J. *J. Chrom.* **1959**, *2*, 490.
(11) Hofstee, M. T.; Kwantes, A.; Rijnders, C. W. A. *Symposium Distillation Brighton* **1960**, 105.
(12) Deal, C. H.; Derr, E. L. *Ind. Eng. Chem., Process Des. Dev.* **1964**, *3*, 394.
(13) Chatterjee, A. K.; King, J. W.; Karger, B. L. *J. Colloid Interface Sci.* **1972**, *41*, 71.
(14) Larkin, J. A.; Pemberton, R. C. *Natl. Lab. (U. K.) Rep.* **1973**, *24*, 163.
(15) Mash, C. J.; Pemberton, R. C. *NPL Report Chem.* July, **1980**, 111.
(16) Barr, R. S.; Newsham, D. M. T. *Fluid Ph. Equil.* **1987**, *35*, 189.
(17) Gautreaux, M. F.; Coates, J. *AIChE J.* **1955**, *1*, 496.
(18) Scott, L. S. *Fluid Ph. Equil.* **1986**, *26*, 149.
(19) Trampe, D. M.; Eckert, C. A. *J. Chem. Eng. Data* **1990**, *35*, 156.
(20) Slocum, E. W.; Dodge, B. F. *AIChE J.* **1964**, *10*, 364.
(21) Tochigi, K.; Kojima, K. *J. Chem. Eng. Japan* **1976**, *9*, 267.
(22) Tochigi, K.; Kojima, K. *J. Chem. Eng. Japan* **1977**, *10*, 343.
(23) Lobien, G. M.; Prausnitz, J. M. *Ind. Eng. Chem. Fundam.* **1982**, *21*, 109.
(24) Bergmann, D. L.; Eckert, C. A. *Fluid Ph. Equil.* **1991**, *63*, 141.
(25) Hassam, A.; Carr, P. W. *Anal. Chem.* **1985**, *57*, 793.
(26) Park, J. H.; Hassam, A.; Couasnom, P.; Frity, D.; Carr, P. W. *Anal. Chem.* **1987**, *59*, 1970.
(27) Abraham, M. H.; Grellier, P. L.; Mana, J. *J. Chem. Thermodyn.* **1974**, *6*, 1175.
(28) Sagert, N. H.; Lau, D. W. P. *J. Chem. Eng. Data* **1986**, *31*, 475.
(29) Dallas, A. J. *Ph. D. Thesis* University of Minnesota: Minneapolis, MN, **1991**.
(30) Leroi, J. C.; Masson, J. C.; Renon, H.; Fabries, J. F.; Sannier, H*Ind. Eng. Chem., Process Des. Dev.* **1977**, *16*, 139.
(31) Richon, D.; Philippe, A.; Renon, H. *Ind. Eng. Chem., Process Des. Dev.* **1980**, *19*, 144.
(32) Richon, D.; Renon, H. *J. Chem. Eng. Data* **1980**, *25*, 59.
(33) Duhem, P.; Vidal, J. *Fluid Ph. Equil.* **1978**, *2*, 231.
(34) Lee, H. J. *Hwahak Konghak* **1983**, *21*, 317.
(35) Lebert, A.; Richon, D. *J. Agric. Food Chem.* **1984**, *32*, 1156.
(36) Richon, D.; Sorrentino, F.; Voilley, A. *Ind. Eng. Chem., Process Des. Dev.* **1985**, *24*, 1160.
(37) Sorrentino, F.; Voilley, A.; Richon, D. *AIChE J.* **1986**, *32*, 1988.
(38) Locke, D. C.; Martire, D. E. *Anal. Chem.* **1967**, *39*, 921.

(39)   Janini, G. M.; Qaddora, L. A. *J. Liq. Chrom.* **1986**, *9*, 39.
(40)   Djerki, R. A.; Laub, R. J. *J. Liq. Chrom.* **1988**, *11*, 585.
(41)   Sorensen, J. M.; Arit, W. *Liquid-Liquid Equilibrium Data Collection, Binanry Systems*; Behrens, D.; Eckermann, R.; Chemistry Data Series; Schon & Wetzel: GMBH, Frankfurt/Main, F.R. Germany, 1979, Vol. 1, Part 1.
(42)   Riddick, J. A.; Bunger, W. B.; Sakaro, T. K. *Organic Properties and Methods of Purification*, Forth Edition; John Wiley & Sons Inc.: New York, NY, 1986; 1042.
(43)   Nicolaides, G. L.; Eckert, C. A. *Ind. Eng. Chem. Fundam.* **1978**, *17*, 331.
(44)   Lafyatis, D. S.; Scott, L. S.; Trampe, D. M.; Eckert, C. A. *Ind. Eng. Chem. Res.* **1989**, *28*,585.
(45)   Howell, W. J.; Karachewski, A. M.; Stephensen, K. M.; Eckert, C. A.; Park, J. H.; Carr, P. W.; Rutan, S. C. *Fluid Ph. Equil.* **1989**, *52*, 151.
(46)   Gmehling, J.; Rasmussen, P.; Fredenslund, A. *Ind. Eng. Chem., Process Des. Dev.* **1982**,*21*, 118.
(47)   Weidlich, U.; Gmehling, J. *Ind. Eng. Chem. Res.* **1987**, *26*, 1372.
(48)   Gmehling, J.; Tiegs, D.; Knipp, U. *Fluid Ph. Equil.* **1990**, *54*, 147.
(49)   Pierotti, G. J.; Deal, C. H.; Derr, E. L. *Ind. Eng. Chem. Fundam.* **1959**, *51*, 95.
(50)   Fujita, T.; Nishioka, T.; Nakajima, M. *J. Medicinal Chem.* **1977**, *20*, 1071.

RECEIVED April 20, 1992

Chapter 17

Polycyclic Aromatic Hydrocarbon Analogues
Another Approach to Environmental Separation Problems

A. Cary McGinnis[1], Srinivasan Devanathan[1], Gabor Patonay[1,3],
and Sidney A. Crow[2]

[1]Department of Chemistry, Georgia State University, Atlanta, GA 30303
[2]Department of Biology, Georgia State University, Atlanta, GA 30303

Aryl dyes containing different aromatic groups may be prepared using
simple synthetic steps, e.g., by condensation of an arylcarboxaldehyde
with a quaternary salt of a heterocycle having a methyl group in the 2-
or 4-position. The chromophore in these dyes, due to the more
extensive conjugation, has much longer absorption and emission
wavelengths than the aryl compound itself allowing the researcher to
utilize the lower interference spectral region, e.g., the red or near-
infrared region. The spectral properties of aryl dyes may be used to
significantly reduce separation needs when the fate of aryl compounds
is followed in the environment. This paper focuses on the spectral
properties and environmental utility of aryl dyes prepared in our
laboratory for decreasing separation requirements.

The worldwide exploration and shipment of crude oil and petroleum products have
led to a number of catastrophic environmental disasters. A wealth of information
exists on these events; however, long-term consequences are poorly understood. In
addition, the processing and utilization of coal and petroleum fuels invariably lead to
low levels of pollution in the terrestrial and marine environment by a variety of
molecules including polycyclic aromatic hydrocarbons (PAHs). Pollution occurs at
different levels which can either be acute or chronic. The effects of the latter are the
most difficult to discern.

The importance of studying the fate of PAHs in the environment can not be
understated. PAHs are a ubiquitous, diverse group of organic compounds containing
one or more fused aromatic rings. PAHs are found in air, water and soil samples due
to contamination from combustion of hydrocarbons and from petroleum spills. But
the fate of PAHs in the environment is hard to follow as a result of detection and

[3]Corresponding author

0097–6156/92/0509–0229$06.00/0
© 1992 American Chemical Society

separation difficulties. Simple spectroscopic and chromatographic methods are severely limited because of interference in biological systems.

Current mechanisms to evaluate the degradation and detoxification of anthropogenic PAHs in the environment frequently depend on complex analytical schemes which involve the extraction of large volumes of material. There are several possible chemical and biological routes through which PAHs are changing in the environment. For example, bacteria and fungi have been known to metabolize polycyclic aromatic hydrocarbons [1,2]. Owing to the extremely complicated chemical and biological processes, a large number of derivatives of PAHs may be found in the environment. This fact may explain the difficulties associated with environmental determination of PAHs. Advancements in separation science are major factors in the recent development of PAH research in the environmental sciences. In spite of the advances in separation sciences, however, there are still major difficulties due to significant interference from other environmentally important compounds [3-5].

The development of appropriate analogues is one possible means of decreasing interference and reducing the need for complicated separation procedures in studies of the environmental fate of PAHs. These analogues can be used for predicting the fate of PAHs in the environment using much simpler analytical procedures than otherwise required. We are in the process of developing an even simpler analytical approach to studying the fate of PAHs in the environment by using red and near infrared analogues of PAHs. Using these compounds, we can simplify the analytical procedure in environmental analyses by moving the absorption wavelengths away from the interference of the sample where the underivatized PAH molecules are normally seen. The direct use of absorption or fluorescence spectroscopy for the study of the biologically and environmentally important PAHs may be less valuable due to interference.

PAHs are spectroscopically active in the region where interference is more likely. These difficulties warrant the investigation of molecules that may serve as model compounds or analogues for studying biological systems and the fate of PAHs in biological systems. These needs led us to the investigation of aryl dyes.

Aryl dyes may be prepared by forming a methine chain between a quaternary salt of a heterocyclic compound having a methyl group in the 2- or 4-position and an aryl compound, e.g. PAHs [6]. The best known compounds in this group are the styryl dyes. Following a synthetic route similar to that used in the preparation of styryl dyes, other dyes containing aryl groups may be prepared. For example, styryl dyes are prepared by the condensation of benzaldehydes with the quaternary salt of heterocyclic compounds [6].

Pyrene is one of the best characterized PAH molecules. Its spectral and chemical properties as well as its role and fate in the environment have been thoroughly investigated [3]. For example, pyrene has been found to be a useful fluorescence probe for studying microenvironmental changes since it is very sensitive to different quenching processes and changes in microhydrophobicity [7-9]. In this manuscript, the spectroscopic characteristics and the interaction with the microbial population of a new aryl dye, will be discussed. The information gained by spectroscopic techniques provides an alternative approach to separation problems.

Materials and Methods

The methanol used in this study was obtained from the Aldrich Chemical Company (Milwaukee, WI) in spectrometric grade. Other chemicals used in the synthesis of the aryl dye were also obtained from Aldrich. The chemical structure of the red absorbing pyrene analogue is given in Figure 1 along with some other aryl dyes that may be easily synthesized from commercially available precursors. Only preparation of the pyrenyl dye is given below because other aryl dyes may be prepared similarly using the appropriate aryl carboxaldehyde [6].

The starting material was 3-ethyl-2-methylbenzothiazolium iodide obtained by heating a solution of ethyl iodide and 2-methylbenzothiazole in dimethylformamide at 150 °C for 12 hours. Workup of this reaction mixture included cooling to 23 °C and slow addition of diethyl ether to crystallize the product. The crystals were filtered, washed with diethyl ether, recrystallized and dried under reduced pressure to give the analytically pure 3-ethyl-2-methylbenzothiazolium iodide, m.p. 241-243 °C. The subsequent transformation of this iodide was as follows: A mixture of 1 g (3.8 mmol) 3-ethyl-2-methylbenzothiazolium iodide, 0.898 g (3.9 mmol) of 1-pyrenecarboxaldehyde and 0.726 g (7.2 mmol) of dry triethylamine was refluxed in 5 ml of ethanol for 30 minutes. The reaction mixture was cooled and the solvent was removed under reduced pressure. The crude solid thus obtained was dissolved in dichloromethane and precipitated with ether and further purified with recrystallization. Chromatographic purification may be necessary if crystals do not form readily. The m.p. of the analytically pure compound is 225-228 °C, decomposing. The structure of the dye, given in Figure 1a, was fully consistent with the measured NMR spectrum (CDCl$_3$). HPLC analysis has been performed for this series of dyes. A chromatogram of the phenanthrenyl dye on a C-8 column is shown in Figure 2. This chromatogram is representative of the aryl dyes and indicated retention times are similar to that of underivatized PAHs.

Instrumentation. Absorption spectra for the pyrenyl dye were obtained by using a Perkin-Elmer Lambda 2 UV/Vis/NIR spectrophotometer. The spectrophotometer is interfaced to a 286 computer to store spectra and control the instrument. Each spectrum was recorded using a PECSS program (Perkin-Elmer). Fluorescence spectra were recorded on an SLM 8000 spectrofluorometer interfaced to an IBM PS/2 computer.

The 10^{-3} M stock solutions of the pyrenyl dye were prepared in spectrophotometric grade methanol or ethanol. For the studies, solutions were prepared by pipetting the required amount of dye stock solution dissolved in alcohol into a volumetric flask which was filled with water or culture media. The preparation of dye solutions has been found to be much easier when this method is followed instead of completely removing the methanol from the aliquot of the stock solution by nitrogen gas. The final solutions contained a negligible amount of methanol which did not have any effect on the results.

In the process of developing this analytical approach to studying the environmental fate of PAHs using red and NIR absorbing aryl dyes as analogues, we

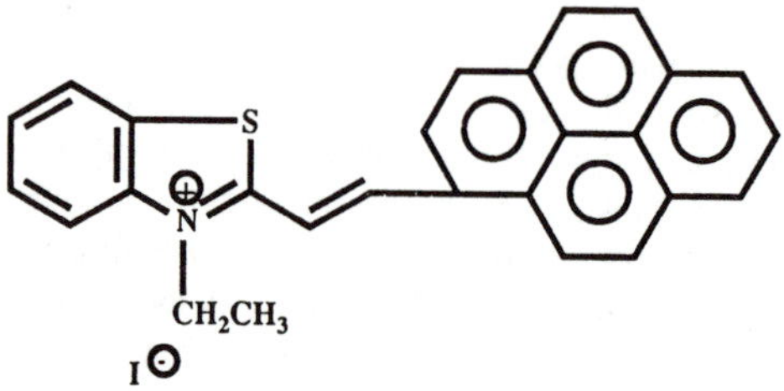

Figure 1a. Chemical structure of the red absorbing pyrenyl dye used in this study.

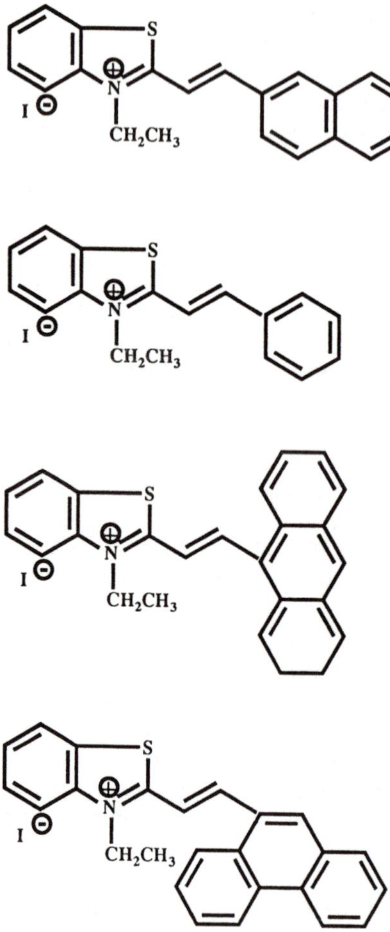

Figure 1b. Chemical structure of different aryl dyes.

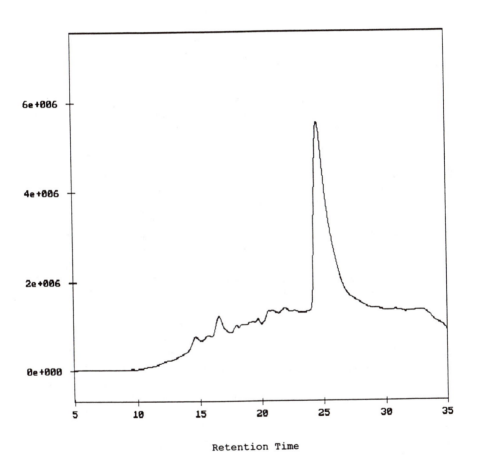

Retention Time

Figure 2.   HPLC chromatogram of phenanthrenyl dye on C-8 column (gradient: 15%-80% isopropanol in water, both being 5mM in pentane sulfonic acid).

expect to significantly reduce the separation requirements during environmental analysis. By using these aryl dyes, we hope simply to follow metabolitic changes of PAHs in the environment without complicated separation steps. This is achieved by moving the absorption away from the interference of environmental samples where the underivatized PAH molecule is normally seen. The effectiveness of this approach may be evaluated by studying the effect of dye on yeast cultures that are known to metabolize PAHs.

Studies of biological interaction of the PAH analogues

Using the pyrene analogue, preliminary toxicity and binding studies were conducted. Fresh liquid biological cultures were prepared from stock cultures Mycological agar (Difco) for yeasts and Saboraud's Dextrose agar for filamentous fungi by transferring sufficient inoculum to an appropriate broth. Yeasts were routinely cultured on 67% yeast nitrogen base (Difco) supplemented with 0.5% glucose (GYNB). Filamentous fungi were cultured on Saboraud's Dextrose broth (Difco) because of the more complex requirements and slower growth rates. In-depth studies of biological activity were conducted with the yeasts.

Toxicity. Initial assessment of the toxicity of these aryl dye molecules using a modification of the agar disk diffusion antibiotic sensitivity was inconclusive because of the limited diffusion of the compound and its intense binding to the cellulose disks. Liquid cultures supplemented with 1 mg of aryl dye dissolved in 1 ml of DMF were used to assay toxicity. Comparison of the growth of Candida lipolytica (GSU 37-1) and Candida maltosa (GSU R-42) was made by observing the optical density at 595 nm in a Turner spectrophotometer model 380. Determination of the absorption by compound was made for uninoculated GYNB. The increase in absorbance in cultures with and without analogue was compared.

Uptake. Association of a molecule with a non-growing cell in a non-growth supporting environment is an important parameter for biologically active molecules. Because of differences in the absorption spectra of our analogues in solvents of different hydrophobicity, we were interested in seeing the effects in biological systems of effectively different physiological and morphological nature. Cultures were grown under similar basal conditions but supplemented with either 0.5% or 5.0% glucose, 0.5% ethanol, or 0.5% tetradecane. Cultures were inoculated as described above and incubated on an R. Braun gyrotary shaker (autoshake model U) at 120 cycles per minute. Cultures were incubated at room temperature (approximately 24 °C) for appropriate intervals sufficient to produce the necessary cell mass. Following incubation, an aliquot of cells was removed from the culture. The cells were sedimented by centrifugation (10 min at 3000 times G) and washed in 0.9% saline 3 times and resuspended to the original 10 ml. Different concentrations of aryl dye were added to the cell suspension in DMF (.5mg/.5ml). The mixture was allowed to incubate in the dark for 30 min. Following incubation the cells were again removed by centrifugation and the concentration of aryl dye was determined spectrophotometrically on a Perkin Elmer Lambda 2 UV/Vis/NIR spectrophotometer.

Results and Discussion

Representative absorption and fluorescence spectra for the phenanthrenyl dye in methanol and water are shown in Figure 3. Detailed spectroscopic studies of the pyrenyl dye compound in the presence of organized media has been described earlier [10,11]. As can be seen from the data, these dyes can change their absorption spectra depending upon the hydrophobicity of the environment similar to the pyrenyl dye [10,11]. These changes, which have been associated with dimerization [10,11], may result in color changes of the dye solution when the hydrophobicity of the solvent is changing. Since the absorption spectra of the dyes indicate whether the dye molecule is in monomeric or dimeric form, useful information may be gained concerning its interaction with cell membrane.

It is interesting to compare the absorption spectra of these dyes to that of the PAH molecules [10,11]. If we compare the region of the spectra where PAH molecules absorb (250-400 nm), we can observe interesting similarities [10,11]. Figure 4 illustrates these similarities. These indicate that the spectral properties of the aryl group are retained to a great extent. The spectroscopic similarities between these dyes and their underivatized PAH counterparts actually go beyond the ones mentioned above and have been discussed earlier [10,11]. These similarities, however, do not indicate whether we can utilize these compounds as PAH analogues. To evaluate this aspect, we studied the interaction of these dyes with different yeast strains.

Table 1 reflects the growth of cultures with and without the aryl dye following adjustment for the PAH analogue absorption. There is no evidence of growth inhibition in either of these cultures. Furthermore, subsequent studies with the filamentous fungi; Aspergillus niger, Cunninghamella elegans, and Syncephalastrum racemosa have demonstrated no growth inhibition by the analogue. Uptake is distinctly influenced by the growth substrate (Table 2). The greatest specific binding (micrograms/10^6 cells) was with Candida lipolytica grown on ethanol, followed by tetradecane grown cells. Candida maltosa demonstrated a slightly different pattern (greatest binding in low glucose and very similar activity for tetradecane, and ethanol substrates). Least efficient binding for either culture was from high glucose conditions.

These preliminary studies have demonstrated the utility of the study of analogues in discerning biologically significant processes such as toxicity and uptake. The differences noted suggest major differences in the uptake of our analogues as a function of growth substrate and a significant species-dependent component of the initial encounter with these analogues, similar to underivatized PAHs [1,2]. A wider study of these analogues and their interaction with microbial cultures and communities under a variety of conditions will enhance our ability to extrapolate from analogue to PAH effects. Detailed studies comparing the analogue behavior to PAH and studies to determine other facets of the biological effects of PAH are in progress.

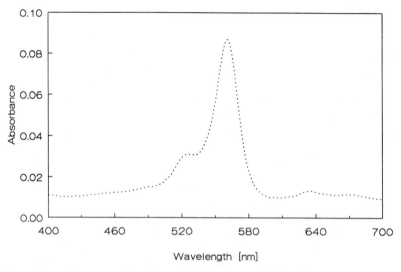

Figure 3a. Representative absorption spectrum of the phenanthrenyl dye (10^{-6} M) in methanol.

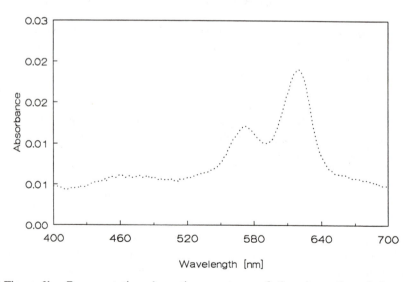

Figure 3b. Representative absorption spectrum of the phenanthrenyl dye (10^{-6} M) in water.

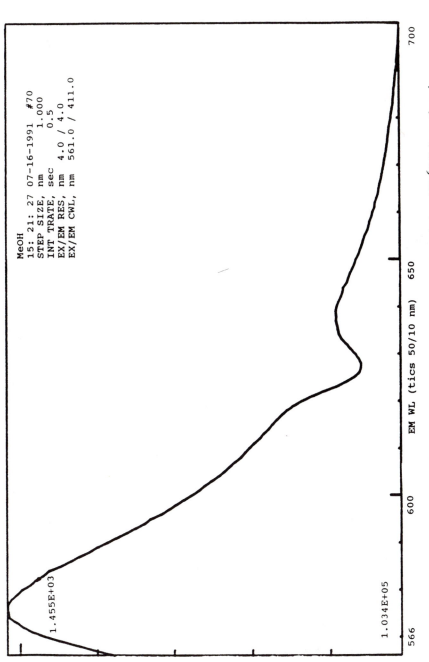

Figure 3c. Representative fluorescence spectrum of the phenanthrenyl dye (10^{-6} M) in methanol.

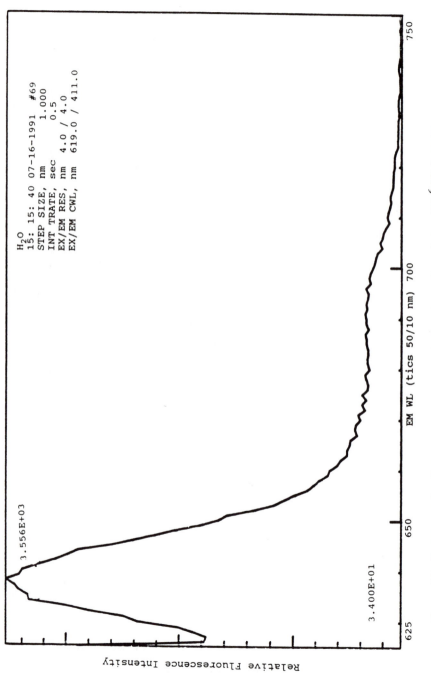

Figure 3d. Representative fluorescence spectrum of the phenanthrenyl dye (10^{-6} M) in water.

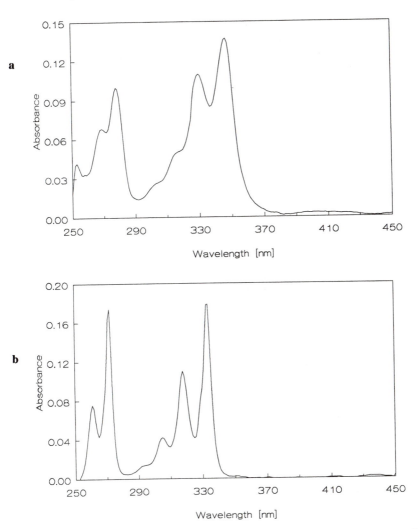

Figure 4. Representative absorption spectra in the UV region of the pyrenyl dye (a) and pyrene (b) in methanol.

Very little or no separation is required during determination of these aryl dyes in the media. Although the media exhibit a broad absorption peak severely interfering with the underivatized aryl compounds, very little interference is observed in the region where the chromophore absorbs, indicating the utility of using these aryl dyes as analogues to study the environmental fate of PAHs.

Syntheses of additional analogues with a different number of fused rings are in progress. A systematic evaluation of the activities of these molecules in biological systems will provide a sound basis for the application of red and near-infrared laser

Table 1
Effect of NIR pyrene-analogue on growth of yeast in 0.5% GYNB

	Candida maltosa (R-42) 0.5%GYNB[1] + analogue		Candida lipolytica (37-1) 0.5%GYNB[1] + analogue	
Time [hours]	Absorption		Absorption	
0	.01*	.01**	.02*	.02**
12	.16	.15	.09	.08
24	.60	.59	.47	.49
48	1.22	1.27	1.08	1.06
96	1.59	1.59	1.74	1.73

[1] 0.5% Glucose in 0.67 Yeast Nitrogen Base (YNB)
* Absorption at 595nm
** Absorption at 595nm corrected for compound absorbance

Table 2
Binding of NIR pyrene-analogue as influenced by culture media

		Candida maltosa (R-42)			Candida lipolytica (37-1)		
Media	Time	% bound	cell #'s	specific binding*	% bound	cell #'s	specific binding*
0.5%TYNB[1]	24h	98	1.2E7	84	94	2.3E6	408
0.5%GYNB[2]	24h	96	8.9E6	107	88	2.8E7	31
5% GYNB[3]	24h	14	2.5E7	6	43	4.7E7	10
0.5%EYNB[4]	24h	92	1.4E7	83	97	1.8E6	532

* micrograms binding for 1E6 cells (approx 1 microgram cells)

[1] 0.5 % Tetradecane in 0.67% Yeast Nitrogen Base (YNB)
[2] 0.5 % Glucose in 0.67% YNB
[3] 5 % Glucose in 0.67 YNB
[4] 0.5 % Ethanol in 0.67 YNB

spectroscopy to the evaluation of analyses where otherwise extensive separation would be required. The major advantage of this method is that the absorption maxima of the analogue may be chosen using appropriate heterocyclic derivatization. Using aryl dyes as analogues, not only may the separation aspect of environmental analyses be simplified, but it may also improve our understanding of microbial detoxification processes in cases where the environmental damage has already occurred.

We have evaluated <u>Candida lipolytica</u> and <u>Candida maltosa</u> as to the utility of using red or near-infrared absorbing aryl dyes to decrease separation requirements in environmental analyses. These, as well as other bacteria and fungi, are well documented as degraders of hydrocarbons and PAHs in marine environments. First aryl dyes were introduced into cell cultures to study the recovery and mass balance of the systems. These measurements may be conducted easily since very little or no separation is required due to significantly decreased spectral interference. Before we can fully utilize aryl dyes as PAH analogues, however, several important points need to be addressed. Further studies are needed to determine the biotic and abiotic degradation routes of these aryl dyes. These studies are presently under way in our laboratory.

Acknowledgement

This work was supported in part by a grant from the National Science Foundation (CHE-890456) and in part under a cooperative agreement from EPA (R818696).

References

1. D.T. Gibson, *Crit. rev. Microbiol.*, **1**, (1972) 199.
2. D.T. Gibson, "Biodegradation of Aromatic Petroleum Hydrocarbons," In, D.A. Wolfe ed., "Fate and Effects of Petroleum Hydrocarbons, Pergamon Press, New York, NY (1977).
3. M.C. Bowman, "Handbook of Carcinogens and Hazardous Substances," Marcel Dekker, New York, NY (1982).
4. H.V. Gelboin, P.O.P. Ts'o, "Polycyclic Hydrocarbons and Cancer, " Academic Press, New York, NY (1978).
5. C.E. Cerniglia and S.A. Crow, *Arch. Microbiol.*, **129**, (1981) 9.
6. Rodd's Chemistry of Carbon Compounds, Vol IV, Part B., Ed: S. Coffey, Elsevier Scientific Publishing Company, (1977), pp. 370-422.
7. G. Patonay, K. Fowler, A. Saphira, G. Nelson and I.M. Warner, *J. Incl. Phenom.*, **5**, (1987) 717.
8. K. Kalyanasundaram and J.K. Thomas, *J. Am. Chem. Soc.*, **99**, (1977) 2039.
9. R.C. Benson, H.A. Kues, *J. of Chem. Eng. Data,* **22**, (1977) 379.
10. M.D. Green, G. Patonay, T. Ndou and I.M. Warner, *J. Incl. Phenom,* in press.
11. A.E. Boyer, S. Devanathan, and G. Patonay, *Anal. Lett.* **24**, (1991) 701.

RECEIVED April 20, 1992

Chapter 18

Trace Element Distribution in Various Phases of Aquatic Systems of the Savannah River Plant

Shingara S. Sandhu

Claflin College, Orangeburg, SC 29115

The distribution of Cd, Cu, Fe, Mn, Ni and Zn species was estimated for the dissolved solid (TSS) phases of thermally impacted and non-impacted aquatic SRP systems. The major fractions of Cd, Cu, Ni and Zn were present as dissolved ions while most of the Fe was present in the solid phase. Dissolved species of Cu, Fe, Mn and Ni were insensitive to natural Ca and alkalinity gradient across SRP aquatic systems, whereas the dissolved species of Cd and Zn and solid phase exchangeable Zn, responded to this gradient. Solid phase Cd was primarily found in the exchangeable and carbonate phase. Zn and Ni did not display a clear distribution pattern between various components of the solid phase. The increase in the percentage of dissolved Cd is accounted for by source water chemistry and thermal conditions associated with cooling water activities. Increased competition between Zn and Ca for exchange sites produced a decrease in solid associated exchangeable Zn.

Elevated concentrations of potentially toxic heavy metals in water and bottom sediments are commonly associated with many industrial processes and human activities. Metals entering the aquatic systems, whether from atmospheric or terrestrial sources, partition into different components of solid and dissolved phases. the bioavilability and toxicity of these metals are not solely a function of their total concentrations, but rather, are related to their partitioning between the solid and solution phases. Even in the solution phase, the relative toxicity of these metals is often related to their chemical forms such as inorganically complexed ions, exchangeable ions, and organically complexed ions. In general, the free ionic form of metals is relatively more toxic than the complexed forms (1) as it tends to interact more readily than other forms which may also be present in solution. It has been suggested (2) that differences in bioavailability of dissolved Pb and Cd reflected their differential tendencies to form complexes with dissolved organic moities. The bioavailability of sediment associated Ag, Cd, Cu, Mn, Pb and Zn to

0097–6156/92/0509–0242$06.00/0

benthic organisms was significantly influenced by the elements distribution in various sediment phases (3,4). Therefore, the knowledge of metal species in the natural system can aid in understanding their behavior and bio-availability.

Though several studies (5-9) have described the concentrations of trace elements in surface waters, most of them did not differentiate between species of a particular element. Some (10) have considered the distribution between dissolved and particulate forms. However, few attempts (11) have been made to evaluate the distribution of metals between various solid phase components of the suspended material. The present study provides quantitative estimates of dissolved (dissolved is defined as those aquatic components that could not be removed by centrifugation from liquid phase) and various solid phase associated metal fractions in southeastern United States streams. Between November' 83 and August ' 85, 46 bi-weekly samples were taken from six SRP associated watersheds to determine dissolved (filterable) and total element concentrations. As one of several goals of the study was to assess the impact of natural and production related activities on trace element behavior in these aquatic systems, knowledge of speciation within solid and dissolved phases was essential for data interpretation. The research described herein used sequential extraction and a thermodynamics approach to define solid and dissolved phase species of Cu, Cd, Fe, Mn, Ni and Zn. The study also evaluated the effects of natural and production related processes on the distribution of metals in aquatic systems at SRP.

MATERIALS AND METHODS
Study Area. The SRP (Figure 1) is a 780 square kilometer nuclear production facility located on the upper Atlantic coastal plain, adjacent to the Savannah River. There is a northwest to southeast transition in the geological features underlying the SRP (12). The study area and its watersheds ,impacted by SRP activities, have been described in an earlier report (13).
Sampling Sites. Three sampling sites ,6,29 and 34, were selected on the Savannah River to characterize the waters used for cooling purposes. Sites 30, 31, and 32 were selected on the upper Three Runs Creek. Site 33 was selected on the Tims Branch which is a tributary to the Upper Three Runs Creek . Sites 27 and 28 were selected on Beaver Dam Creek. Sites 23 and 25 were selected on non-thermal branch of Four Mile Creek. An additional site, 19, was selected on Pen Branch Creek (non-thermal area) which is also a branch of Four Mile Creek. Two sites, 22 and 24, selected on Four Mile Creek and two sites ,18 and 20, selected on Pen Branch Creek were located in thermal areas. Four sites 13, 14,15 and 16, on Steel creek were in post-thermal region. Two of these sites,14 and 16, were in an area of active lake construction and were also receiving some process sewage from the 100P area which used Savannah River water. Sites 10,11 and 12, were located on Meyers Branch which is a subsidiary to Steel Creek. The Meyers Branch never received thermal effluent. Site 4 on Par Pond receives warm water from a pre-cooling pond B which then circulates through the Par Pond

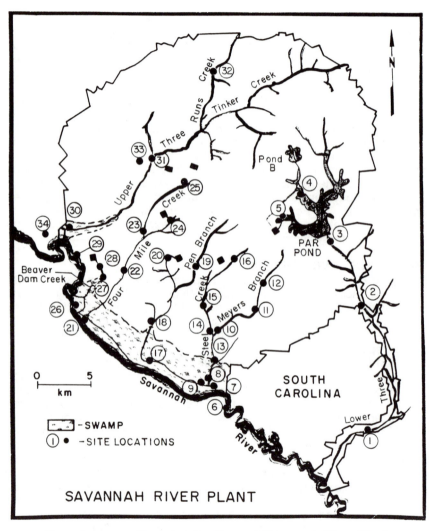

Figure1.   Sampling locations at Savannah River Plant.

and returns to the P-Reactor via an intake at site 5. Some water from Par Pond flows over a dam at site 3, which is located immediately above the head of Lower Three Runs Creek. Finally, two additional sites (1 and 2),were established on Lower Three Runs Creek and six sites,7, 8,9,17,21 and 26, were established in the river swamp below the regions of entry for the various creeks.

Background Water Chemistry. Fifty-five variables were measured from November 1983 to August 1985 for 46 bi-weekly water samples, collected from 34 sites (13). The variables used in statistical identification of the factors controlling water quality in the SRP aquatic systems were the following: stream velocity, water temperature, percentage of oxygen saturation, pH, chloride, total alkalinity, total suspended solids, sulfate, dissolved ortho-phosphate, particulate-associated phosphorus, nitrate, ammonia, organic nitrogen, dissolved calcium, dissolved magnesium, dissolved potassium and dissolved sodium.   All analyses were performed under quality control/quality assurance conditions specified by the South Carolina Department of Health and Environmental Control (13).

Speciation Sampling. Twenty-five liter surface water samples were collected from the middle of each creek, pond ,etc. in linear polyethylene bottles during May 28/29, 1985 and June 11/12, 1985. Roughly one half of the sites were sampled during each period with the restriction that all samples within a watershed were taken on the same date. An aliquot of each sample was then filtered through a 0.45 µm Millipore type PC filter that had been soaked in 0.05 N HCl for 4 hours and thoroughly rinsed with double-deionized water.   The filterates were received in 500 mL, acid-cleaned Teflon bottles, acidified with Ultrex nitric acid to a pH of 2.0 and stored at 4°C for dissolved metal analysis. Procedural blanks were generated by passing double-deionized water through this process.   All glassware and plasticware were soaked (50% v/v) in concentrated nitric acid for at least 2 days and then thoroughly rinsed with double-deionized water.   The reagents used for the study were of analytical grade or better; each was tested for potential contamination before use.   Metals measured in filterates were defined as dissolved metals.

Total suspended solids (TSS) were collected from 15 liters of each sample with a Sorvall SS-3 centrifuge, equipped with a "Szent-Gyorgyi and Blum" KSB continuous flow system. Titanium tubes were used to minimize contamination. The system was operated at 10,000 rpm with a flow rate of 125 mL/min. The solids were quantitatively transferred to 50 mL, polycarbonate centrifuge tubes and excess water was removed by centrifugation. Solids were frozen and later extracted in these centrifuge tubes.   Additional material to determine total suspended solids (TSS) was collected from 8 liters of water by use of the continuous flow system. Finally the solids were transferred to porcelain crucibles and dried at 110° C to a constant weight.

Dissolved Phase Metals.   Fe and Mn were analyzed directly in filterates by flameless atomic absorption spectrophotometery.   One hundred mL of filterate was used for the concentration of Cd, Cu, Ni and

Zn by using the cobalt dithiocarbamate co-precipitation and back extraction (14) method. The blanks underwent a similar procedure. Both blanks and sample extracts were analyzed with a Hitachi 180-80 atomic absorption spectrophotometer, equipped with Zeeman background correction. The method of standard additions was used to correct matrix effects during analysis. The analytical conditions during the experiment were those recommended by the vendor (15).

Solid Phase Metals. Trace metals in the solid phase were sequentially extracted for (I) exchangeable, (II) bound to carbonates, (III) bound to oxides of iron and manganese, (IV) bound to organic matter, by using the available method (16) which is based on the empirical understanding of the system rather than on any strong theoretical or mathematical concepts. These methods can be imprecise due to a degree of extractant non-selectivity and redistribution among phases during extraction (17). However; soil scientists have been using a similar approach for meeting the nutritional needs of plants for decades. Subsequent studies (18,19) suggest that the efficiency and selectivity of this (16) method may be acceptable for certain systems, as highly significant correlations were found between some trace metals and iron and Mn extracted with a reducing agent. It is suggested that the designations of exchangeable, bound to carbonates, bound to iron and manganese oxides, and bound to organic matter, be considered as procedural definitions that reflect the relative distributions of element species in the SRP aquatic systems.

To minimize the loss of materials, selective extractions were conducted in the centrifuge tubes in which the solids were originally stored. The volumes of the extracting solutions were proportioned to the weight of solids in the tube. Between each successive extraction, separation was affected by centrifugation (Sorvall Model RC2-B) at 10,000 rpm for 30 minutes. The supernatant was removed with a pipet and analyzed for trace elements. The concentrations of these elements in the extracts were determined by flameless atomic absorption spectrophotometry as described for the dissolved trace element analyses (15). The total concentration of a solid phase-associated trace element is the sum of the four extraction concentrations and does not include the residual (14) or crystalline (11) solid fraction.

RESULTS AND DISCUSSION

Total Suspended Solids. The easily releasable trace elements are generally enriched in the endogenic material whose particulates are usually very small and are often found deposited on the clastic material (20). The total suspended solids separated by continuous flow centrifugation system from 8 liters of water were the composites of clastic and edogenic materials. There was a wide range of variation in the TSS(96.2 mgL^{-1} for Steel Creek Sample 16 to 0.30 mgL^{-1} for Swamp Sample 17) with a median value of 13.5 mgL^{-1}. The total suspended solids concentrations were high in the portion of the Steel Creek Samples (14 and 16) due to reservoir building activities in the area at the time of sampling.

The total suspended solids in the water systems under study were low when compared to U.S. rivers (21) where the mean suspended load was 602 mgL^{-1}. However, the values for TSS reported for this study are comparable to the previously published information (10,22) for SRP and river Verkaan in Sweden (23) which showed an average of 10.2 mgL^{-1} of TSS. The Upper Three Runs and their tributaries raised the suspended load of Savannah River from 8.82 mgL^{-1} at site 34 to 9.96 mgL^{-1} at site 29. However, the suspended load of Savannah river dropped to 5.32 mgL^{-1} by the time its water reached site 6.   The rapid drop in the concentration of TSS between the sites may have to do with the Chemistry of the Savannah River water whose water conductivity was about four times (85.0) higher than the Upper Three Runs Creek.   The ionic concentration changes in the immediate vicinity of particulates in an aqueous system are expected to result in flocculation, coagulation, aggregation and finally to the sedimentation of the suspended materials. The activities at the SRP did not contribute significantly to the suspended load of the Savannah River water.

DISTRIBUTION OF ELEMENTS: The sampling scheme was chosen to elucidate the distribution of trace metals and to study the influence of several variables on the distribution and chemistry of metals in the SRP aquatic systems.   The dissolved element concentrations found in the present study were similar to those reported previously (10,22) for some of the sites.   The results of analyses for Cu and Fe as fractions of the total element concentration are given in Figures 2 and 3.   The range and median for all elements are given in Table I.

The addition of Savannah river water to the SRP water systems imparted a chemical signature to receiving waters. Generally, chloride, sulfate, magnesium, sodium and potassium concentrations of the receiving streams or reservoirs increased relative to nonimpacted areas (13).   In addition, the chemistry of the non-thermally impacted water systems was influenced by the groundwater as it led to an increase in Ca and alkalinity concentrations in watersheds, fed by groundwater.

The Tuscaloosa formation contributed significant amounts of very soft water to the Upper Three Runs Creek watershed.   The remaining watersheds received significant water input from the McBean formation. The yield and calcareous nature of the McBean formation groundwater increased with its distance southeast-wards (12); consequently, the systems receiving groundwater from this formation displayed increased Ca and alkalinity in a seaward direction.

The data (Table I, Figures 2 and 3) clearly show that the majorities of the Cu , Cd, Zn, and Ni were present in the dissolved phase; most of the Fe was found in the solid phase.   Dissolved and solid phases were equally important for Mn content in these systems.   These trends were consistent with the   transport of metals in watersheds in eastern Tennessee (24), Maryland (25), and Australia (26) as well as earlier work (27) on the Upper Three Runs Creek and Steel Creek watersheds. However; this data does not support the previous work (11) on the transport of metals by the Amazon and Yukon Rivers where most of the Fe as well as the other trace elements were associated with the various components of the solid phase.   These differences may be resolved by considering the TSS load which was much higher in the Amazon and

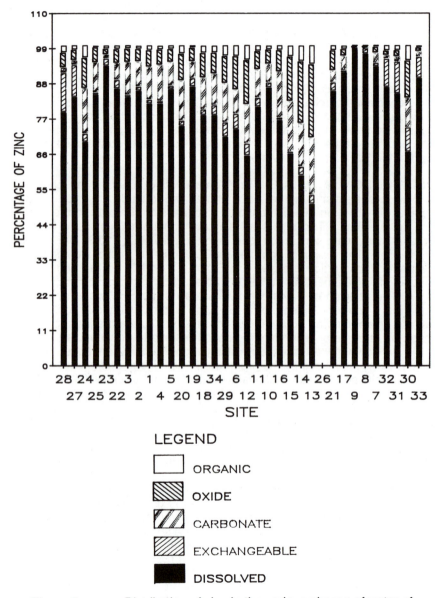

Figure 2.   Distribution of zinc in the various phases of water of
                 SRP.

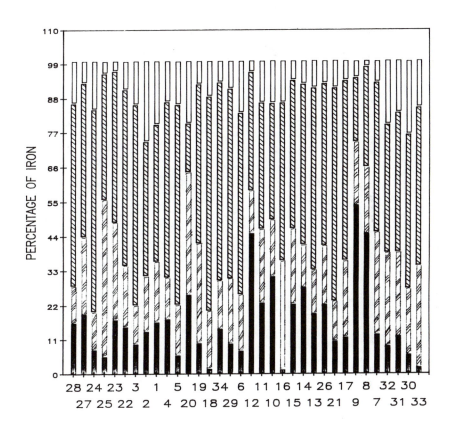

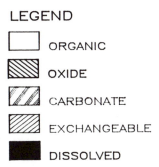

Figure 3.   Distribution of iron in the various phases of water of SRP.

TABLE I. Summary of Metal Concentrations and Speciation in SRP Water Systems

SPECIES	Cu		Cd		Zn		Ni		Fe*		Mn	
	Med.	Range	Med.	Range	Med.	Range	Med.	Range	Med.	Range	Med.	Range
Solid Phase												
Exchangeable%	1	<1-22	36	1-71	15	2-67	11	1-79	<1	<1	44	7-77
Carbonate%	20	9-77	36	6-96	41	7-63	25	6-45	25	13-53	25	8-51
Oxide %	12	3-24	5	<1-27	30	11-48	38	8-82	61	21-76	27	9-63
Organic%	66	5-82	8	<1-82	11	3-36	18	4-43	11	4-30	3	1-12
Total (mg/Kg)	20	3.3-504.9	2.6	0.4-9.3	76	10-279	13	3-268	24	10-95	3.4	0.2-22.3
Dissolved Phase %												
Free Ion	-	-	75	17-87	48	6-68	40	2-63			79	32-89
MSO_4	-	-	<1	<1-2	<1-1	<1-1	-	-			<1	<1-2
MCl^+	-	-	<1	<1-2	1-	-	-	-			-	-
$MHCO_3^+$	-	-	<1	<1-3	<1	<1-2	<1	<1-1			-	-
$MCO_3(AQ)$	-	-	<1	<1-56	<1	1-10	15	1-94			-	-
M Fulvate	>99	52-99	19	<1-67	48	9-89	41	2-85			20	10-68
$M(OH)^+$	-	-	<1	<1-7	<1	<1-26	-	-			<1	<1-7
$M(OH)_2 (AQ)$	<1	1-48	<1	<1-7	<1	<1-50	<1	<1-1			-	-
Dissolved Phase		0.3-		.02-		1.3-		1.3-				
Total (ug/L)	0.7	5.9	0.14	2.72	3.1	4.5	2.3	12.3	32	4-293	32	5-7
$K_d(10^3)$	301	1.9-504.9	17.2	1.6-387.8	25.3	3.3-79.0	6.0	0.8-109.9	366	36-10.644	82	15-1437

*$Fe(OH)^+_2$ = 90% Range 21-98, $Fe(OH)_3$ (AQ) 9%, Range 1-34 $Fe(OH)^-_4$ <1% Range 1-89

Yukon Rivers. The results for Mn concentrations in SRP watersheds were inconsistent with the other studies (27,28) where the characteristics of the aquatic systems were markedly different from that of the present systems. Gibbs (28) examined a large river with high TSS loadings and Giesy et. al. (27) focussed on a subset of the nonimpacted sites, included in the present study. Concentrations of dissolved metals were consistently higher in SRP watersheds relative to those of the Walker Branch watershed in eastern Tennessee (24) but concentrations of dissolved Cu and Cd in SRP aquatic systems were similar to Norris Brook watersheds (29). The low concentrations of Cd, Cu, Mn, Fe and Zn in the Walker Branch watershed were attributed to factors such as (1) the relatively undisturbed condition of the catchment, (2) the unmineralized nature of the underlying soils, and (3) low solubilities of oxide, carbonate and hydroxide components of the solid phase due to relatively high pH (7.9) and well oxygenated conditions of watershed.

The DOC concentrations at baseflow conditions noted in Walker Creek (24) were approximately 0.2 mg L^{-1}. In contrast, the DOC concentrations in the SRP systems ranged from 1.3 to 16.7 mg L^{-1} and those in Norris Brook (29) were approximately 1 to 5 mgL^{-1}. The laboratory studies (30) conducted on the speciation of metals, using SRP watershed waters, suggested that low pH and high DOC content of these coastal plain water systems enhanced the solubility and subsequent complexation of various solid phase associated metal species. Higher concentration of dissolved Cd, Cu, Mn and Fe observed in Norris Brook aquatic system (29) relative to Walker Branch (24) may be attriluted to low pH (5.7 -6.4) and high DOC concentrations of Norris Brook water. The pH changes in addition to controling the precipitation and dissolution of solid phase associated metals may also significantly affect the surface reactions between the metals and substrate (31) which could have contributed to the lack of agreement between the results from various watersheds.

Speciation estimation for the elements in dissolved phase was performed with the MINTEQ thermodynamic equilibrium model (32). This model was modified (33) by use of stability constants for Ni, Cd, Cu, Mn, Mg, and Ca complexes with fulvate. The modified model has been accepted and applied (30, 34) to the various systems at SRP as well as elsewhere (35,36). It was assumed for the application of this model to SRP data that all dissolved organic carbon characteristics were reflected in the fulvate fraction. These assumptions render the results of this exercise semiquantitative. However; the information necessary for more precise estimates is not available for naturally occurring organic ligands (37, 38). It was difficult to resolve for the application of this model, whether Fe in the filterate was present in dissolved or colloidal forms. Therefore, the model was executed once without allowing precipitation and, again, allowing ferrihydrite and goethite precipitation.

The data for the speciation of dissolved metals are given Table I. Major fractions of dissolved Cu, Cd, Zn, Ni and Mn were present as fulvate complexes or free ions. The data reported here are consistent with the earlier studies (27) for the Upper Three Runs Creek, and the Bever Dam Creek. Small fractions of these metals were predicted to be present as $M(OH)^+$, $M(OH)_2(aq)$, $M(HCO_3)^+$ and $MCO_3(AQ)$. The model predicted a very small fraction of Cd, Zn and Mn present as MSO_4. Fe was predicted by this model to be present in the form of various hydroxides.

Although there were considerable variations in the distribution of solid phase-associated metals between sites, some general trends were apparent. Copper associated with the TSS was present predominantly in the organic fraction (13). In contrast, the highest percentages of solid phase Cd were associated with the exchangeable and carbonate fractions (13). Zn and Ni tended to have higher percentages associated with the carbonate and oxide fractions (Fig.2). Highest percentages of Mn were found in the exchangeable fraction although the fractions of Mn found in carbonate and oxide fractions were also important. Fe was predominantly present in oxide component of the solid phase (Fig. 3).

STATISTICAL CORRELATIONS: A statistical evaluation of the analytical results, using the Pearson Product-moment correlation (39) was performed to obtain information about elements that co-existed in the different components of a particular phase. The correlation analysis of six elements for 34 sampling stations is given in Tables II and III which shows that the TSS correlated positively with oxide and organic Cd species and exchangeable Mn. All species of Fe and Zn showed strong, positive correlations with TSS (Table II.) This supports the observation, that a greater percentage of metals might have been associated with the solid phase if the levels of TSS in the SRP water systems were higher than the levels found at the time of this study.

The role of Cu and Ni appears to be different from previously mentioned elements as only organic species of these elements showed positive correlation with TSS (Table II.) The behavior of Cu and Ni which showed their independence from TSS, was difficult to explain. However, it is clear that these two elements are more selective in their association with TSS as they were preferably incorporated into organic matter (Table II) as compared to other elements, which appeared to have been incorporated predominantly in the inorganic fraction of the TSS. This was especially true for Fe where a major fraction of solid phase-associated iron was present as oxide and all of the dissolved phase Fe was predicted by the model to be present as hydroxides (Table I).

The dissolved Cu showed strong positive correlation with dissolved and exchangeable Mn whereas Cu and Mn were independent of Fe concentrations. It can be concluded from this observation that the elements Cu and Mn were independent of the scavenging effects of Fe. Only 2.8 percent of Cu and 10.9 percent of Mn were associated with the oxide fraction of solid phase as compared to 5.35 percent of Cu and 14.49 percent of Mn which were found to be associated with the carbonate fraction of the solid phase(13). It has been reported that generally, in the surface layers of natural waters, the conditions of pH and oxygen are such that Mn^{2+} introduced into these layers is readily oxidized to insoluble MnO_2 (40). This apparently was not true for the surface waters at SRP.   The average pH of water samples examined at

TABLE II. Total Suspend Solids Correlations with Species of Elements

SPECIES	ELEMENTS					
	Cu	Fe	Mn	Zn	Mi	Cd
Dissolved	0.30	0.70	-0.00	0.34	-0.08	-0.15
Exchangeable	-0.13	0.88	0.78	0.21	0.02	0.24
Carbonate	-0.07	0.42	0.26	0.82	0.20	-0.07
Oxide	0.19	0.87	0.25	0.90	0.13	0.39
Organic	0.43	0.62	0.11	0.84	0.97	0.87

TABLE III. Intra Species Correlations

ELEMENT	SPECIES	DISS.	EXCH.	CARB.	OXID.	ORG.
				Fe		
	Diss.	1.00	0.68	0.27	0.63	0.27
	Exch.	0.68	1.00	0.18	0.64	0.38
Fe	Carb.	0.27	0.18	1.00	0.73	0.61
	Oxid.	0.63	0.64	0.73	1.00	0.77
	Org.	0.27	0.38	0.61	0.77	1.00
				Mn		
	Diss.	1.00	0.23	0.38	0.39	0.11
	Exch.	0.23	1.00	0.62	0.43	0.22
Mn	Carb.	0.38	0.62	1.00	0.80	0.67
	Oxid.	0.39	0.48	0.80	1.00	0.85
	Org.	0.11	0.22	0.67	0.85	1.00
				Zn		
	Diss.	1.00	0.17	0.48	0.44	0.44
	Exch.	0.17	1.00	0.01	0.22	0.34
Zn	Carb.	0.43	0.01	1.00	0.93	0.89
	Oxid.	0.45	0.22	0.93	1.00	0.97
	Org.	0.44	0.34	0.89	0.97	1.00
				Cu		
	Dis..	1.00	0.22	0.05	0.14	0.20
	Exch.	0.22	1.00	0.25	0.44	0.04
Cu	Carb.	0.04	0.24	1.00	0.39	0.22
	Oxid.	0.14	0.44	0.39	1.00	0.26
	Org.	0.20	0.04	0.22	0.26	1.0

SRP was close to 6.96 (13), which probably favored the formation of $MnCO_3$ rather than MnO_2 even though the waters were well aerated (minimum level of dissolved oxygen observed was 4.7 mgL^{-1}).

The Savannah River water impacted the composition of SRP watersheds which received effluents from the power generating facilities by decreasing the amounts of exchangeable Mn, Ni, and Zn and increasing the amounts of oxide and carbonate forms of these elements. A representative of data is given in Figure 4. Turbidity, colloidal ortho-phosphate phosphorus, particulate aluminum, and particulate Fe probably contributed to the amount of TSS. All these variables were strongly influenced by the water turbulence regime. TSS and other TSS associated variables were responsible for 17 percent of variance in the SRP water systems (13). Total alkalinity, dissolved calcium, total organic carbon and dissolved Fe were another set of variables which were responsible for 15 percent of variance within the SRP water systems (13). This implies that the transition in the calcareous nature of the geological features (12) beneath the SRP was an important factor which contributed to the variations in the water quality at the SRP. The dissolved Zn decreased whereas the carbonate Zn increased from NW to SE (Fig. 5) transition of the underlying calcareous nature of the geological material at SRP. Calcium and total alkalinity in SRP surface waters increased with greater water yield and calcareous nature of the groundwater with distance southeastwards across the SRP (12).

The Upper Three Runs Creek is located at the extreme NW whereas the Lower Three Runs Creek is located at the extreme SE of the SRP (Fig. 1). There was a slow but consistent change in the concentration of exchangeable Zn which decreased in creeks going from NW to SE and carbonate Zn which increased in the creeks as one goes across from NW toward a SE direction at SRP. There appears to be a mild positive correlation (Table III) between exchangeable Zn and dissolved Zn. However, the concentration of dissolved Zn had to exceed a certain minimum for it to affect the increase in the concentration of exchangeable Zn. This was particularly true as long as the concentration of Ca in water systems remained low. It is apparent that with rising Ca concentration, the alkalinity of water systems also interjected into this equation. The solubility product constant (Ksp) for Zn CO_3 in neutral aqueous systems at 25°C is 8×10^{-5}. Conceptually, the concentration of dissolved Zn in the aqueous system has to increase beyond its Ksp values for it to be associated with carbonate, presuming all other variables remained favorable for this reaction . Maximum concentration of Zn observed at the SRP water systems is 4.47 μgL^{-1} which is below the threshhold value required for the formations of Zn CO_3. This shows that relation between Zn^{2+} and CO_3^{2-} was controlled by pH of water systems rather than the threshhold values of Zn.

The temperature and percent oxygen saturation were associated with the thermal characteristics of sampling sites and were responsible for 14 percent of variance in the SRP water systems. The thermal regime significantly increased the concentration of Mg, Na and ortho-phosphate

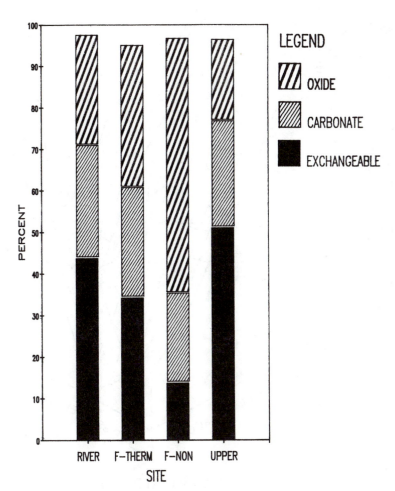

Figure 4.   Impact of Savannah River water and Thermal conditions on solid phase manganese.

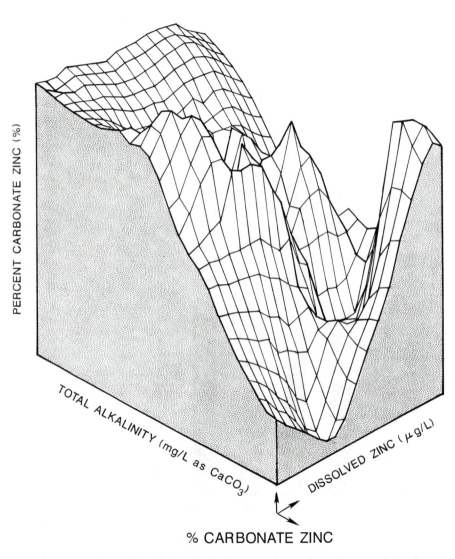

% CARBONATE ZINC

Figure 5.   Effect of total alkalinity on the carbonate associate zinc.

in the SRP water systems (13). It also slightly increased the total alkalinity (13).

DYNAMICS OF DISTRIBUTION: The natural aqueous system is a complex multiphase system which contains dissolved chemicals as well as suspended solids. The metals present in such a system are likely to distribute themselves between the various components of the solid phase and the liquid phase. Such a distribution may attain (a) a true equilibrium or (b) follow a steady state condition. If an element in a system has attained a true equilibrium, the ratio of element concentrations in two phases (solid/liquid), in principle, must remain unchanged at any given temperature. The mathematical relation of metal concentrations in these two phases is governed by the Nernst distribution law (41) commonly called the partition coefficient (K_d) and is defined as $K_d = a_{(s)} / a_{(l)}$ where a(s) is the activity of metal ions associated with the solid phase and $a_{(l)}$ is the activity of metal ions associated with the liquid phase (dissolved). This behavior of element is a direct consequence of the dynamics of ionic distribution in a multiphase system. For dilute solution, which generally obeys Raoult's law (41) activity (a) of a metal ion can be substituted by its concentration, (c) moles L^{-1} or moles Kg^{-1}. This ratio (K_d) serves as a comparison for relative affinity of metal ions for various components-exchangeable, carbonate, oxide, organic-of the solid phase. Chemical potential which is a function of several variables controls the numerical values of K_{ds} (41).

Partition co-efficients (K_{ds}) for all species of each element (K_d exchangeable, K_d carbonate, K_d oxides, K_d organic) along with total K_d (K_d total mass) were calculated for 34 sampling sites. The representative values for total K_{ds} are given in Table I. The partition co-efficients (K_{ds}) reported here are similar to those previously published (10). It appears that the nature and the chemical make up of the TSS are more important for their interaction with the metal ions than their total mass as the sample which contained 1.34 mgL^{-1} of TSS had a K_d of 2.0×10^4 and a sample which contained 96 mgL^{-1} of TSS had a K_d value of 1.0×10^4. There are wide variations in the values of partition coeffeients. Fe which showed the highest values for K_d (Table I) also had the highest amount of oxide-associated concentration. Interestingly, dissolved (filterable) Fe was predicted by the MINTEQ thermodynamic equilibrium model to be present entirely in the form of hydroxides of iron which can be interpreted as an evidence for the suspension of Fe precipitates which may have acted as substrates for surface adsorption reaction in the aquatic systems. Mn appears to have followed similar pattern.

The K_d value for each site is large and varies from every other sampling location. This leads to the hypothesis that each sampling site though connected through waterways has its own salient characteristics and acted like an independent entity. The large value of K_d for each sampling site indicates that the cumulative effect of multi-site adsorption (surface exchange), surface reaction and precipitation led to the formation of stable endogenic products. There was no likelihood for the reverse reaction which can bring the metal ions back into solution, to occur as long as the internal and the external forces (pH, redox, temperature etc.) governing the system remain unchanged.

The use of the partition coefficient (K_d) has been considered questionable by some workers (10). However, to validate such use of the partition coefficient (K_d) one must realize that its values must be based on thermodynamic equilbrium of the system which is unlikely to happen in the natural aquatic system due to frequent fluctuations in temperature, pressure ,pH,Eh, etc. The true thermodynamics equilibrium is sensitive to the internal and external stresses and relates to a limited number of concentrations encountered in the reaction system. Theoretical considerations dictate that the partition coefficient (K_d) computations, based on true thermodynamic equilibrium, remain constant as long as the other variables such as temperature, pH, Eh, pressure, etc. remain unchanged.

Fe, Mn and Zn, which are more sensitive to pH and redox changes and have low solubility product constants as compared to Cu and Ni (42) appeared to have maintained strong intra-species positive correlations (Table III). On the other hand, Cu and Ni were observed to have behaved differently in the SRP aquatic systems as they did not show intra-species positive correlations. The dissolved (filterable) species of these elements (Cu and Ni) in the water systems were conceived to have acted independently from the rest of their species-exchangeable, carbonate, oxide and organic.

The aquatic systems at SRP, like any other system, contained several types of reacting and adsorbing ligands. Theromdynamic considerations require that the metal ions distribution in these systems continue to move towards achieving equilibria with each component of the aqueous and the solid phases. The driving force for partitioning the metal ions into various fractions is chemical potential. As long as the system remains at equilibrium, the chemical potential of the system remains constant and the system remains unperturbed. When the equilibrium controlling variables like temperature, pressure, pH, Eh etc. change due to internal or external stresses (man made or natural), the chemical potential of the system is unbalanced which drives the system back to its new equilibrium condition.

CONCLUSIONS

The chemical characteristics of the SRP water systems receiving Savannah River water and or thermal effluents from the power generating facilities have changed. Elevated concentrations of chloride, sulfate, magnesium, sodium, potassium, manganese, cadmium and zinc were observed in the impacted watersheds. The data suggest that each sampling site, though linked through waterways showed its independent characteristics.

Normal activities of SRP did not significantly affect the physical and chemical characteristics of the Savannah River water.

ACKNOWLEDGEMENTS

This research was supported by contract DE-AC09-76ROO-819 between the U.S. Department of Energy and SREL Aiken, South Carolina. I sincerely thank Dr. Micheal C. Newman for his support at every phase of this work.

Literature Cited

1. Kennaga, E.E. and Goring, C.I. In *Aquatic Toxicology*; Eaton, J.G.; Parrish, P.R. and Hendricks, A.C. (Eds.); American Society forTesting and Materials: Philadelphia, PA, 1980.
2. Wiener, J.G. and Giesy, J.P. Jr., Concentrations of Cd, Cu, Mn, Pb and Zn in a highly organic softwater pond. Fish. Res. Board Can; 1979, 36, 270-279.
3. Ray S., McLeese, D. W. and Peterson, M.R., Accumulation of copper, zinc, cadmium and lead from two contaminated sediments by three marine invertebrates - A laboratory study. Bull. Environ. Contam. Toxicol., 1981, 26, 315-322.
4. Thomson, E. A. Luoma, S. N. Johansson, C. E. and Cain, D. J. Comparison of sediments and organisms in identifying sources of biologically available trace metal contamination. Water res., 1984, 18, 755-765.
5. Dresser, D.R., Gladney, E.S., Owens, J.W.; Perkins, B.L., Winke, C.L. and Wangen, L.E., Comparasion of trace elements extracted from fly ash levels found in efflunt waters from coal fired powerplant. Envion. Sci. Technol. ,1977, 11, 1017.
6. Cherry, D.S. and Guthrie, R.K., Mode of elemental dissipation from ash basin effluent. Water, Air, Soil Pollut.,1978, 9. 403.
7. Chu, T. J., Ruane, R. J. and Krenken, P.A. J., Characterization and reuse of ash pond effluents in a coal-fired power plant. Water Pollut. cont. Fed., 1978, 21, 2494-2498.
8. Evans, D.W., Alberts,J.J. and Clark III, R.A., Reversible ion exchange fixation of cesium-137 leading to mobilization from reservoir sediments. Geochem. Cosmochem. Acta., 1983, 47, 1041-1048.
9. Kopp, J. F. and Kroner, R.C. A comparison of trace elements in natural waters; Dissolved verses suspended, In *Developments in applied spectroscopy*, Baer, W.K. (Ed), Plenum Press, New York, 1968, 6, 339.
10. Alberts, J.J., Newman, M.C.and Evans, D.W., Seasonal variations of trace elements in dissolved and suspended loads for coal ash ponds and pond effluents. Mechanisms of trace metal transport in rivers. Water, Air, Soil Pollut., 1985, 26, 111-28.
11. Gibbs,R.J., Mechanisms of trace metal transport in rivers Science,1973,180,71-3.

12. Siple, G.E., "Geology and groundwater of the Savannah River
     Plant and vicinity" South Carolina Geological Survey,Water
     Supply 1967 ,Paper 1841.
13. Newman, M.C., Dancewicz A., Davis B., Anderson K., Bayer R.,
     Lew R., Mealy R., Sandhu S.S., Presnell S. and Know J.N.
     Comprehensive cooling water report, Vol. 2 Water Quality.
     Savannah River Ecology Laboratory, Aiken, S.C.,1986, pp. 600.
14. Boyle, E.A. and Edmond, J.M., Determination of trace metals in
     aqueous solutions by APDC chelate co-precipitaion. Analytical
     chemica Acta., 1977, 91, 189-94.
15. Hitachi Applications and Software Manuals, NSI Hitachi Scientific
     Instruments, Mountain View, CA: 1982, 20-25.
16. Tessier, A., Campbell, P.G.C. and Bisson, M., Sequential
     extraction procedure for the speciation of particulate trace metals.
     Anal. chem., 1979; 51, 844-54.
17. Kheboian, C. and Bauer C.F., Accuracy of selective extraction
     procedures for metal speciation in model aquatic sediments., Anal.
     Chem., 1987; 59, 1417-25.
18. Luoma, S.N. and Bryan, and G.W., Factors controlling the
     availability of sediment bound lead. Sci. Total Environ., 1981, 17,
     165-74.
19. Tessier, A. and Campbell , P.G.C., Partitioning of trace metals in
     sediments In metal speciation, In *Theory, analysis and
     application*, J.R. Kramer and H. E. Allen (Eds), Lewis Publisher,
     Inc. Chelsa, MI. 1988.
20. Jenne, E.A., Trace elements sorption by sedirment and soil sites
     and processes. In Molybdenum in the environment:Chapell, W.;
     Peterson, K.K. (Eds). Marcel Dekker Inc. New York, 1977, 2, 42.
21. Judson, S. and Ritter, D.F., J. Geophys Res., 1964, 69, 3395.
22. Evans, D.W. and Giesy, Jr., J.P. In *ecology and coal resources
     development*. Wali, M.K. (Ed) Proc. Intern. Cong. Energy and
     Ecosystem. Grand Fork, N.D. 1978, 2, 782.
23. Carsexuel, L., Mineral equilibria in ecosystem geochemistry. Ecol.
     Bull., 1983, 35, 73.
24. Turner, R. R., Lindberg, S. E. and Talbot, K., Dynamics of trace
     element export from a deciduous watershed, Walker Branch,
     Tennessee, In *Watershed Research in Eastern North America Vol.
     II*, D. L. Correll (Ed.), Smithsonian Institution, Maryland , 1977 pp
     924.
25. Wu, T.L. and Hoopes, M.T. Land utilization and metal discharge
     from the Rhode River watershed. In *Watershed Research in
     Eastern North America, Vol. II*, D.L. Correll (Ed.), Smithsonian
     Institution, Maryland, 1977, pp 924.
26. Hart, B. T., Davies, S.H.R. and Thomas, P.A., Transport of iron,
     manganese, cadmium, copper and zinc by Magela Creek,
     Northern Territory, Australia. Water Res., 1982, 16, 605-12.
27. Giesy, J.P. and Briese, L.A., Trace metal transport by particulates
     and organic carbon in two South Carolina streams. Verh. Int.
     Verein. Limnol., 1978, 20, 1401-1417.

28.   Gibb, R.J., Amazon River: Environmental factors that control its dissolved and suspended load. Science, 1967, 156, 1734-37.

29.   Hall, R.J., Likens G.E., Fiance S. B. and Hendrey, G.R. Experimental acidification of a stream in the Hubbard Brook Experimental Forest, New Hampshire. Ecology 1980, 61, 976-989.

30.   Alberts, J. J. and Giesy J.P., Conditional stability constants of trace metals and naturally occurring humic materials: Application in equilibrium models and verification with field data. In *Aquatic and Terrestrial Humic Materials*, R.F. Christman and E.T. Gjessing (Eds.),Ann Arbor Science Publishers, Ann Arbor, MI, 1983, pp538.

31.   Christensen, T.H., Cadmium soil sorption at low concentration: I. Effect of time, Cadmium load, pH and Calcium. Water, Air, Soil Pollut., 1984, 21, 105-114.

32.   Felmy, A.R., Girvin, D.C. and Jenne, E.A., MINTEQ - A computer program for calculating aqueous qeochemical equilibria. Batelle, Pacific Northwest Laboratories, Richland, WA, 1983, pp 84.

33.   Bertsch, P.M., Loehle, C., Mills, G. and Elrashidi, M.A., Comprehensive Cooling Water Study - Final Report: Heavy Metal Transport. SREL-25 UC66e. National Technical Information Service, Springfield, VA, 1986, pp 95.

34.   Mantura, R.F.C., Organo-metallic interactions in natural waters. In Marine Organic Chemistry, E.K. Duursma and R. Dawson (Eds.), Elsevier Scientific Pub. Co., Amsterdam, 1981, pp 481-86.

35.   Mantura, R.F.C., Dickson, A. and Riley, J.P., The complexation of metals with humic materials in natural waters. Estuarine Coastal Mar. Sci., 1978, 6, 387-408.

36.   Reuter, J.H. & Perdue, E.M., Importance of heavy metal-organic matter interactions in natural waters. Geochim. Cosmochim. Acta .,1977, 41, 325-34.

37.   Malcolm, R.L., Geochemistry of stream fulvic and humic substances. In *Humic Substances in Soil, Sediments and Water*, G.R. Aiken, D.M. McKnight, R.L. Wershaw and P. MacCarthy (Eds.), John Wiley & Sons, New York, 1985 pp. 875.

38.   Johnson,N.M., Mineral equilibria in ecosystem geochemistry. Ecology,1971,52,529-31.

39.   Helwig, J.T. and Council, K.A., SAS Institute MC. Raleigh N.C., 1979, pp 123.

40.   Stum, W. and Morgan, J.J., *Aquatic Chemistry*, Wiley Interscience; New York, 1970, pp. 5-70 and 514-563.

41.   Castillan, G.W., *Physical Chemistry*, Third Ed., Addison-Wesley Publishing Company, Reading, Mass., 1983, pp 307-313.

42.   Christian, G.D., *Analytical Chemistry*, Fourth Ed., John Wiley and Sons, New York, N.Y. , 1986, pp 150-170 and 644-646.

RECEIVED April 3, 1992

INDEXES

Author Index

Affiliation Index

Subject Index

Production: Donna Lucas
Indexing: Deborah H. Steiner
Acquisition: Anne Wilson

Printed and bound by Maple Press, York, PA

Bestsellers from ACS Books

The ACS Style Guide: A Manual for Authors and Editors
Edited by Janet S. Dodd
264 pp; clothbound, ISBN 0–8412–0917–0; paperback, ISBN 0–8412–0943–X

Chemical Activities and Chemical Activities: Teacher Edition
By Christie L. Borgford and Lee R. Summerlin
330 pp; spiralbound, ISBN 0–8412–1417–4; teacher ed. ISBN 0–8412–1416–6

Chemical Demonstrations: A Sourcebook for Teachers,
Volumes 1 and 2, Second Edition
Volume 1 by Lee R. Summerlin and James L. Ealy, Jr.;
Vol. 1, 198 pp; spiralbound, ISBN 0–8412–1481–6;
Volume 2 by Lee R. Summerlin, Christie L. Borgford, and Julie B. Ealy
Vol. 2, 234 pp; spiralbound, ISBN 0–8412–1535–9

Writing the Laboratory Notebook
By Howard M. Kanare
145 pp; clothbound, ISBN 0–8412–0906–5; paperback, ISBN 0–8412–0933–2

Developing a Chemical Hygiene Plan
By Jay A. Young, Warren K. Kingsley, and George H. Wahl, Jr.
paperback, ISBN 0–8412–1876–5

Introduction to Microwave Sample Preparation: Theory and Practice
Edited by H. M. Kingston and Lois B. Jassie
263 pp; clothbound, ISBN 0–8412–1450–6

Principles of Environmental Sampling
Edited by Lawrence H. Keith
ACS Professional Reference Book; 458 pp;
clothbound; ISBN 0–8412–1173–6; paperback, ISBN 0–8412–1437–9

Biotechnology and Materials Science: Chemistry for the Future
Edited by Mary L. Good (Jacqueline K. Barton, Associate Editor)
135 pp; clothbound, ISBN 0–8412–1472–7; paperback, ISBN 0–8412–1473–5

Personal Computers for Scientists: A Byte at a Time
By Glenn I. Ouchi
276 pp; clothbound, ISBN 0–8412–1000–4; paperback, ISBN 0–8412–1001–2

Polymers in Aqueous Media: Performance Through Association
Edited by J. Edward Glass
Advances in Chemistry Series 223; 575 pp;
clothbound, ISBN 0–8412–1548–0

For further information and a free catalog of ACS books, contact:
American Chemical Society
Distribution Office, Department 225
1155 16th Street, NW, Washington, DC 20036
Telephone 800–227–5558